Modelling of computer and communication systems

24 Cambridge Computer Science Texts

Modelling of computer and communication systems

I. Mitrani

Computing Laboratory, University of Newcastle upon Tyne

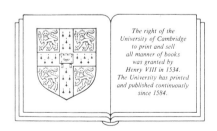

The right of the
University of Cambridge
to print and sell
all manner of books
was granted by
Henry VIII in 1534.
The University has printed
and published continuously
since 1584.

Cambridge University Press

Cambridge

New York   New Rochelle   Melbourne   Sydney

Published by the Press Syndicate of the University of Cambridge
The Pitt Building, Trumpington Street, Cambridge CB2 1RP
32 East 57th Street, New York, NY 10022, USA
10 Stamford Road, Oakleigh, Melbourne 3166, Australia

First published 1987

Printed in Great Britain at the University Press, Cambridge

British Library cataloguing in publication data

Mitrani, I.
    Modelling of computer and communication
    systems. – (Cambridge computer science
    texts; 24)

    1. Data transmission systems. 2. Computer
    networks
    I. Title
    004.6   TK5105.5

Library of Congress cataloguing in publication data

Mitrani, I. (Israel), 1943–
    Modelling of computer and communication systems.

    (Cambridge computer science texts; 24)
    Bibliography
    Includes index.
    1. Digital computer simulation.   2. Telecommunication
systems—Mathematical models.   I. Title.
II. Series.
QA76.9.C65M56 1987   001.4'34   87-877

ISBN 0 521 30688 4 hard covers
ISBN 0 521 31422 4 paperback

MP

Contents

Preface

The designers and users of complex systems obviously have a vested interest in knowing how those systems perform under different conditions. This is no less true of computers and communication networks than it is of motorway junctions, manufacturing facilities and nuclear power stations. In all cases, it is necessary to have a clear understanding of the factors that influence performance and the manner in which they do so. Such understanding can be gained by experimentation – either with the real system, or with a simulated version of it. Alternatively, one can construct a mathematical model of the system and obtain the desired information by analysis. That is the approach that we shall adopt and pursue in this book.

The analytical method has the drawback that the model which is being studied is only an approximation of reality: a number of simplifying assumptions are usually made. However, that need not diminish the value of the insights that are gained. A mathematical model can capture all the essential features of a system, display underlying trends and provide quantitative relations between input parameters and performance characteristics. Moreover, analysis is cheap, whereas experimentation is expensive. A few simple calculations carried out on the back of an envelope can often yield as much information as hours of observations or simulations.

The systems in which we are interested are subjected to demands of random character. The processes that take place in response to those demands are therefore also random. Accordingly, the modelling tools that are needed to study such systems are provided by the theory of probability and stochastic processes. These tools are introduced in chapters 1 and 2. A number of models and results involving single-server and multiple-server systems, with single or multiple job types, are presented in chapters 3 and 4. Some optimisation problems are also examined there. Chapter 5 deals with the important topic of queueing networks, and their applications to multiprogrammed computer systems and computer networks. Finally, chapter 6 is devoted to the modelling and evaluation of certain packet-switching communication networks.

Throughout the text, an effort has been made to make the material easily accessible. There is much emphasis on explaining ideas and providing intuition along with the formal derivations and proofs. Some of the more difficult results are, in fact, stated without proofs. Generality is sometimes sacrificed to clarity. For instance, the mean value treatment of closed queueing networks was chosen in preference to an approach based on the product form solution because it is simpler, albeit somewhat less powerful.

The book is intended for computer science and operations research undergraduates and postgraduates, and for practitioners in the field. Some mathematical background is assumed, including first-year calculus. Readers familiar with probability and stochastic processes may wish to skip some, or all, of chapters 1 and 2.

I would like to thank the individuals and institutions who have helped me to complete this undertaking. Professor Harry Whitfield gave me encouragement and made sure that I did not fall too far behind schedule. Dr Terry Betteridge (University of Newcastle), Dr Albert Greenberg (AT&T Bell Laboratories) and Dr Peter Harrison (University College, London), read the manuscript and contributed a number of helpful comments. Carol Reynolds expertly transformed my scrawl into machine readable form. Last but not least, the Computing Laboratory of the University of Newcastle and the Mathematics Centre of AT&T Bell Laboratories at Murray Hill provided me with pleasant and stimulating environments in which to work.

1
Introduction to probability theory

Unpredictability and non-determinism are all around us. The future behaviour of any system – from an elementary particle to a complex organism – may follow a number of possible paths. Some of these paths may be more likely than others, but none is absolutely certain. Such unpredictable behaviour, and the phenomena that cause it, are called 'random'. Whether randomness is in the nature of reality, or is the result of imperfect knowledge, is a philosophical question which need not concern us here. More important is to learn how to deal with randomness, how to quantify it and take it into account, so as to be able to plan and make rational choices in the face of uncertainty.

The theory of probabilities was developed with this object in view. Its domain of applications, which was originally confined mainly to various games of chance, now extends over most scientific and engineering disciplines.

1.1   Sample points, events and probabilities

We start by introducing the notion of a 'random experiment'. Any action, or sequence of actions, which may have more than one possible outcome, can be considered as a random experiment. The set of all possible outcomes – usually denoted by Ω – is called the 'sample space' of the experiment. The individual outcomes, or elements of Ω, are called 'sample points'.

The sample space may be a finite, denumerable, or non-denumerable set. Its membership depends very much on the definition of 'outcome'.

Examples

1. A race takes place between n horses, with the sole object of determining the winner. There are n possible outcomes, so the sample space can be identified with the set $\Omega = \{1, 2, \ldots, n\}$.

2. In the same race, the finishing position of every horse is of interest and is recorded. The possible outcomes are now the $n!$ permutations of the integers $\{1, 2, \ldots, n\}$ (assuming that two horses cannot arrive at the finishing line at exactly the same time): $\Omega = \{(1, 2, \ldots, n), \ldots, (n, n-1, \ldots, 1)\}$.

3. A certain task is carried out by an unreliable machine which may break down before completion. If that happens, the machine is repaired and the task is restarted from the beginning. The experiment ends when the task is successfully completed; the only output produced is the number of times that the task was started. The sample space here is the set of all positive integers: $\Omega = \{1, 2, \ldots\}$.

4. For the same machine, the experiment consists of measuring the period between one repair and the next breakdown. The sample points now are the positive real numbers: $\Omega = \{x; \, x > 0\}$.

The next important concept is that of an 'event'. Intuitively, we associate the occurrence of an event with certain outcomes of the experiment. For instance, in example 1, the event 'an even-numbered horse wins the race' is associated with the sample points $\{2, 4, 6, \ldots, n - (n \bmod 2)\}$. In example 3, the event 'the task is run unsuccessfully at least k times' is represented by the sample points $\{k+1, k+2, \ldots\}$. In general, an event is defined as a subset of the sample space Ω. Given such an event, A, the two statements 'A occurs' and 'the outcome of the experiment is one of the points in A', have the same meaning.

The usual operations on sets – complement, union and intersection – have simple interpretations in terms of occurrence of events. If A is an event, then the complement of A with respect to Ω (i.e. those points in Ω which are not in A) occurs when A does not, and vice versa. Clearly, that complement should also be an event. It is denoted by *not* A, or A^c. If A and B are two events, then the union of A and B (the sample points which belong to either A, or B, or both) is an event which occurs when either A occurs, or B occurs, or both. That event is denoted by A *or* B, or $A \cup B$. Similarly, the intersection of A and B (the sample points which belong to both A and B) is an event which occurs when both A and B occur. It is denoted by A *and* B, or $A \cap B$, or A, B, or AB.

There is considerable freedom in deciding which subsets of Ω are to be called 'events' and which are not. It is necessary, however, that the definition should be such that the above operations on events can be carried out. More precisely, the set of all events, $\mathscr{A}$, must satisfy the following three axioms.

E1: The entire sample space, Ω, is an event (this event occurs no matter what the outcome of the experiment).

E2: If A is an event, then A^c is also an event.

E3: If $\{A_1, A_2, \ldots\}$ is a finite or denumerable set of events, then the union

$$A = \bigcup_i A_i$$

is also an event.

From E1 and E2 it follows that the empty set, $\varnothing$, is an event (that event can never occur). From E2 and E3 it follows that if $\{B_1, B_2, \ldots\}$ is a finite or denumerable set of events, then the intersection

$$B = \bigcap_i B_i = \left(\bigcup_i B_i^c\right)^c$$

is also an event.

In set theory, a family which satisfies the above axioms is called a σ-field (or a σ-algebra, or a Borel field). Thus the set of all events, $\mathscr{A}$, must be a σ-field. At one extreme, $\mathscr{A}$ could consist of Ω and $\varnothing$ only; at the other, $\mathscr{A}$ could contain every subset of Ω.

Two events are said to be 'disjoint' or 'mutually exclusive' if they cannot occur together, i.e. if their intersection is empty. More than two events are disjoint if every pair of events among them are disjoint. A set of events $\{A_1, A_2, \ldots\}$ is said to be 'complete', or to be a 'partition of Ω', if (i) those events are mutually exclusive and (ii) their union is Ω. In other words, no matter what the outcome of the experiment, one and only one of those events occurs.

To illustrate these definitions, consider example 3, where a task is given to an unreliable machine to be carried out. Here we can define $\mathscr{A}$ as the set of all subsets of Ω. Two disjoint events are, for instance, $A = \{1, 2, 3\}$ (the task is completed in no more than three runs) and $B = \{5, 6\}$ (it takes five or six runs). However, if we include, say, event $C = \{6, 7, \ldots\}$ (the task needs at least six runs to complete) then the three events A, B, C are not disjoint because B and C are not. The events A and C, together with $D = \{4, 5\}$, form a partition of Ω.

Having defined the events that may occur as a result of an experiment, it is desirable to measure the relative likelihoods of those occurrences. This is done by assigning to each event, A, a number, called the 'probability' of that event and denoted by $P(A)$. By convention, these numbers are normalised so that the probability of an event which is certain to occur is 1 and the probability of an event which cannot possibly occur is 0. The probabilities of all events are in the (closed) interval $[0, 1]$. Moreover, since the probability is, in some sense, a measure of the event, it should have the additive property of measures: just as the area of the union of non-intersecting regions is equal to the sum of their areas, so the probability of the union of disjoint events is equal to the sum of their probabilities.

Thus, probability is a function, P, defined over the set of all events, whose values are real numbers. That function satisfies the following three axioms.

P1: $0 \leqslant P(A) \leqslant 1$ for all $A \in \mathscr{A}$.

P2: $P(\Omega) = 1$.

P3: If $A_1, A_2, \ldots$ is a (finite or denumerable) set of *disjoint* events, then

$$P\left(\bigcup_i A_i\right) = \sum_i P(A_i).$$

Note that Ω is not necessarily the only event which has a probability of 1. For instance, consider an experiment where a true die is tossed infinitely many times. We shall see later that the probability of the event 'a 6 will appear at least once', is 1. Yet that event is not equal to Ω, because there are outcomes for which it does not occur. In general, if A is an event whose probability is 1, then A is said to occur 'almost certainly'.

An immediate consequence of P2 and P3 is that, if $\{A_1, A_2, \ldots\}$ is a partition of Ω, then

$$\sum_i P(A_i) = 1. \tag{1}$$

In particular, for every event $A, P(A^c) = 1 - P(A)$. Hence, the probability of the empty event is zero: $P(\varnothing) = 1 - P(\Omega) = 0$. Again, it should be pointed out that this is not necessarily the only event with probability 0. In the die-tossing experiment mentioned above, the probability of the event '6 never appears' is 0, yet that event may occur.

It is quite easy to construct a probability function when the sample space is finite or denumerable. It suffices to assign to the ith outcome a non-negative weight, p_i $(i = 1, 2, \ldots)$, so that

$$\sum_i p_i = 1.$$

Then the probability of any event can be defined as the sum of the weights of its constituent sample points. This definition clearly satisfies axioms P1–P3. Consider again example 3: one possibility is to assign to sample point $\{i\}$ weight $1/2^i$. The events mentioned above, $A = \{1, 2, 3\}$,

$B = \{5, 6\}$, $C = \{6, 7, \ldots\}$ and $D = \{4, 5\}$ would then have probabilities $P(A) = 7/8$, $P(B) = 3/64$, $P(C) = 1/32$ and $P(D) = 3/32$ respectively. Note that the probabilities of A, C and D (those three events form a partition of Ω) do indeed sum up to 1.

When the sample space is non-denumerable (like the positive real axis in example 4), it is more difficult to give useful definitions of both events and probabilities. To treat that topic properly would involve a considerable excursion into measure theory, which is outside the scope of this book. Suffice to say that the events are the measurable subsets of Ω and the probability function is a measure defined over those subsets. We shall assume that such a probability function is given.

If A and B are two arbitrary events, then

$$P(A \cup B) = P(A) + P(A^c B). \tag{2}$$

This is a consequence of the set identity $A \cup B = A \cup (A^c B)$, plus the fact that A and A^c are disjoint. Also, from $B = (AB) \cup (A^c B)$ it follows that $P(B) = P(AB) + P(A^c B)$. Hence,

$$P(A \cup B) = P(A) + P(B) - P(AB). \tag{3}$$

In general, if $A_1, A_2, \ldots$ are arbitrary events, then

$$P\left(\bigcup_i A_i\right) = P(A_1) + P(A_1^c A_2) + P(A_1^c A_2^c A_3) + \cdots \leqslant \sum_i P(A_i). \tag{4}$$

The inequality in (4) becomes an equality only when $A_1, A_2, \ldots$ are disjoint.

The probability of the intersection of two events is not necessarily equal to the product of their probabilities. If, however, that happens to be true, then the two events are said to be independent of each other. Thus, A and B are independent if

$$P(AB) = P(A)P(B). \tag{5}$$

As an illustration, take example 2, where n horses race and there are $n!$ possible outcomes. Let $\mathscr{A}$ be the set of all subsets of Ω and the probability function be generated by assigning to each of the $n!$ outcomes probability $1/n!$ (i.e. assume that all outcomes are equally likely). Suppose that $n = 3$ and consider the following events:

$\quad A = \{(1, 2, 3), (1, 3, 2)\}$ (horse 1 wins);
$\quad B = \{(1, 2, 3), (2, 1, 3), (2, 3, 1)\}$ (horse 2 finishes before horse 3);
$\quad C = \{(1, 2, 3), (1, 3, 2), (2, 1, 3)\}$ (horse 1 finishes before horse 3).

Then events A and B are independent of each other, since $P(AB) = P(\{(1, 2, 3)\}) = 1/6$, and $P(A)P(B) = (2/6)(3/6) = 1/6$. However, events A

and C are dependent, because $P(AC) = P(\{(1, 2, 3), (1, 3, 2)\}) = 2/6$, while $P(A)P(C) = (2/6)(3/6) = 1/6$. It is equally easy to verify that events B and C are dependent.

The above definition of independence reflects the intuitive idea that two events are independent of each other if the occurrence of one does not influence the likelihood of the occurrence of the other. That definition is extended recursively to arbitrary finite sets of events as follows: the n events $A_1, A_2, \ldots, A_n$ ($n > 2$) are said to be 'mutually independent', if

(i) $P(A_1 A_2 \cdots A_n) = P(A_1)P(A_2) \cdots P(A_n)$, and

(ii) every $n - 1$ events among them are mutually independent.

It should be emphasised that neither (i) by itself, nor (ii) by itself, is sufficient for mutual independence. In particular, it is possible that independence holds for every pair of events, yet does not hold for sets of three or more events.

The concepts of independence and dependence are closely related to that of 'conditional probability'. If A and B are two events, then the conditional probability of A, given B, is denoted by $P(A|B)$ and is defined as

$$P(A|B) = P(AB)/P(B). \tag{6}$$

If A and B are independent, then $P(A|B) = P(A)$, which is consistent with the idea that the occurrence of B does not influence the probability of A.

An intuitive justification of the definition (6) can be given by interpreting the probability of an event as the frequency with which that event occurs when the experiment is performed a large number of times. The ratio $P(AB)/P(B)$ can then be interpreted as the frequency of occurrence of AB among those experiments in which B occurs. Hence, that ratio is the probability that A occurs, given that B has occurred.

In terms of conditional probabilities, the joint probability that A and B occur can be expressed, according to (6), as

$$P(AB) = P(A|B)P(B) = P(B|A)P(A). \tag{7}$$

This formula generalises easily to more than two events:

$$P(A_1 A_2 \cdots A_n)$$
$$= P(A_1|A_2 \cdots A_n)P(A_2|A_3 \cdots A_n) \cdots P(A_{n-1}|A_n)P(A_n). \tag{8}$$

The probability of a given event, A, can often be determined by 'conditioning' it upon the occurrence of one of several other events. Let $B_1, B_2, \ldots$ be a complete set of events, i.e. a partition of Ω. Any event, A, can be represented as

$$A = A\Omega = A \bigcup_i B_i = \bigcup_i AB_i,$$

where the events AB_i $(i = 1, 2, \ldots)$ are disjoint. Hence,

$$P(A) = \sum_i P(AB_i) = \sum_i P(A|B_i)P(B_i). \tag{9}$$

This expression is known as the 'complete probability formula'. It yields the probability of A, assuming that those of B_i are known and the conditional probabilities $P(A|B_i)$ can be obtained. We shall see numerous applications of this approach.

Alternatively, having observed that A has occurred, one may ask what is the probability of occurrence of some B_i. This is given by what is known as the 'Bayes formula':

$$P(B_i|A) = P(B_i A)/P(A)$$

$$= P(A|B_i)P(B_i) \bigg/ \left[\sum_j P(A|B_j)P(B_j) \right]. \tag{10}$$

From now on, even when the fact not mentioned explicitly, we shall work in the context of some sample space Ω, set of events $\mathscr{A}$ and probability function P. Such a triple, $(\Omega, \mathscr{A}, P)$, is called a 'probability space'.

Exercises

1. Imagine an experiment consisting of tossing a coin infinitely many times. The possible outcomes are infinite sequences of 'heads' or 'tails'. Show that, with a suitable representation of outcomes, the sample space Ω is equivalent to the closed interval $[0, 1]$.

2. For the same experiment, the event 'a head appears for the first time on the ith toss of the coin' is represented by a sub-interval of $[0, 1]$. Which sub-interval?

3. An experiment consists of attempting to compile three student programs. Each program is either accepted by the compiler as valid, or is rejected. Describe the sample space Ω. Assuming that each outcome is equally likely, find the probabilities of the following events:

      A: programs 1 and 2 are accepted;
      B: at least one of the programs 2 and 3 is accepted;
      C: at least one of the three programs is rejected;
      D: program 3 is rejected.

4. For the same experiment, show that the events A and B are dependent, as are also B and C, and C and D. However, events A and D are independent. Find the conditional probabilities $P(A|B)$, $P(B|A)$, $P(C|B)$ and $P(D|C)$.

1.2   Random variables

It is often desirable to associate various numerical values with the outcomes of an experiment, whether those outcomes are themselves numeric or not. In other words it is of interest to consider functions which are defined on a sample space Ω and whose values are real numbers. Such functions are called 'random variables'. The term 'random' refers, of course, to the fact that the value of the function is not known before the experiment is performed. After that, there is a single outcome and hence a known value. The latter is called a 'realisation', or an 'instance' of the random variable.

Examples

1. A life insurance company keeps the information that it has on its customers in a large database. Suppose that a customer is selected at random. An outcome of this experiment is a collection, c, of data items describing the particular customer. The following functions of c are random variables:

$X(c) = $ 'year of birth';
$Y(c) = $ '0 if single, 1 if married';
$Z(c) = $ 'sum insured';
$V(c) = $ 'yearly premium'.

2. The amount of rainfall over Newcastle upon Tyne during the month of June is recorded. The sample points are now real numbers: $\Omega = \{x; x \geq 0\}$. Those points themselves can be the values of a random variable: $Y(x) = x$.

3. The execution times, x_i, of n consecutive jobs submitted to a computer are recorded. This is an experiment whose outcomes are vectors, v, with n non-negative elements: $v = (x_1, x_2, \ldots, x_n)$; $x_i \geq 0$, $i = 1, 2, \ldots, n$. Among the random variables which may be of interest in this connection are:

$X(v) = \max(x_1, x_2, \ldots, x_n)$ (largest execution time);
$Y(v) = (x_1 + x_2 + \cdots + x_n)/n$ (observed average execution time).

4. A function which takes a fixed value, no matter what the outcome of the experiment, e.g. $X(\omega) = 5$ for all $\omega \in \Omega$, is also a random variable, despite the fact that it is not really 'random'.

Consider now an arbitrary random variable, $X(\omega)$, defined on a sample space Ω, and let (a, b) be an interval on the real line. The set, A, of sample points for which $X(\omega) \in (a, b)$ is the inverse image of (a, b) in Ω. That set represents the event 'X takes a value in the interval (a, b)'. If A is indeed an event, i.e. if $A \in \mathscr{A}$, then the probability $P(A) = P(X \in (a, b))$ is defined. We shall assume that to be the case for all intervals, finite or infinite, open or closed. It is sufficient, in fact, to require that the inverse images of all intervals of the type $(-\infty, x]$ must be events; the rest follows from axioms E1–E3.

The probability that the random variable X takes a value which does not exceed a given number, x, is a function of x. That function, which we shall usually denote by $F(x)$, is called the 'cumulative distribution function', or just the 'distribution function', of X:

$$F(x) = P(X \leqslant x); \quad -\infty < x < \infty. \tag{11}$$

The distribution function of any random variable has the following properties.

1. If $x \leqslant y$ then $F(x) \leqslant F(y)$. This follows from the fact that the event $X \leqslant x$ is included in the event $X \leqslant y$.
2. $F(-\infty) = 0$ and $F(\infty) = 1$. This is because the event $X \leqslant -\infty$ is empty and the event $X \leqslant \infty$ is the entire sample space Ω.
3. $F(x)$ is continuous from the right, i.e. if $x_1, x_2, \ldots$ is a decreasing sequence converging to x, then $F(x) = \lim F(x_i)$. This last property is less obvious than the other two. The idea of its proof is outlined in exercise 1.

Let a and b be two real numbers, such that $a < b$. Since the events $\{X \leqslant a\}$ and $\{a < X \leqslant b\}$ are disjoint, and their union is the event $\{X \leqslant b\}$, we can write $P(X \leqslant a) + P(a < X \leqslant b) = P(X \leqslant b)$. Hence, the probability that X takes a value in the interval $(a, b]$ is given by

$$P(a < X \leqslant b) = F(b) - F(a). \tag{12}$$

If we let $a \to b$ in this equation, we get

$$P(X = b) = F(b) - F(b^-). \tag{13}$$

where $F(b^-)$ is the limit of $F(x)$ from the left, at point b.

Two obvious, but useful, consequences of (12) and (13) are that (i) if $F(a) = F(b)$, then the probability of X taking a value in the interval $(a, b]$ is 0, and (ii) if $F(x)$ is continuous at point b, then the probability of X being equal to b is 0. Remember that an event may have probability 0 without being impossible.

Figure 1.1 illustrates two important types of distribution functions.

Figure 1.1

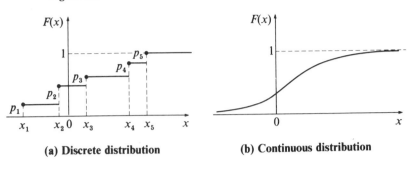

(a) Discrete distribution   **(b) Continuous distribution**

The function exemplified in part (a) of figure 1.1 is constant everywhere, except at a finite or denumerable set of points $x_1, x_2, \ldots$, where it has jumps of magnitudes $p_1, p_2, \ldots$ respectively ($p_1 + p_2 + \cdots = 1$). A random variable with such a distribution function can take the value x_1 with probability p_1, x_2 with probability p_2, etc.; the probability of observing it anywhere else is 0. Random variables of this type, and their distribution functions, are said to be 'discrete'.

From now on, when we talk about the 'distribution' of a discrete random variable, that term will usually refer to the values x_i and their probabilities, p_i ($i = 1, 2, \ldots$), rather than the function $F(x)$.

The random variables in examples 1 and 4 of this section are discrete.

The distribution function in figure 1.1(b) is continuous everywhere. A corresponding random variable can be observed in any interval where $F(x)$ has a non-zero increment, with non-zero probability. However, the probability that it takes any particular value is zero. Such random variables and distribution functions are called 'continuous'.

The random variables in examples 2 and 3 are continuous.

In the case of continuous random variables, the derivative of the distribution function plays an important role. That derivative is called the 'probability density function', or PDF, of the random variable. We shall denote it usually by $f(x)$. According to the definition of derivative, we can write

$$f(x) = \lim_{\Delta x \to 0} \frac{F(x + \Delta x) - F(x)}{\Delta x}$$

$$= \lim_{\Delta x \to 0} \frac{P(x < X \leqslant x + \Delta x)}{\Delta x}. \tag{14}$$

Hence, when Δx is small, $f(x)\Delta x$ is approximately equal to the probability that the random variable takes a value in the interval

$(x, x + \Delta x]$. This is what the term 'density' refers to. In what follows, we shall sometimes write that the probability that the random variable takes value x, rather than being a straightforward 0, is given by

$$P(X = x) = f(x)dx, \tag{15}$$

where dx is an infinitesimal quantity. This, of course, is nothing but a shorthand notation for equation (14).

A distribution function, even when it is continuous everywhere, is not necessarily differentiable everywhere. However, in all cases of interest to us, the set of points where the PDF does not exist will be finite in any finite interval.

Given a probability density function $f(x)$, the corresponding distribution function is obtained from

$$F(x) = \int_{-\infty}^{x} f(u)du. \tag{16}$$

In order that $F(\infty) = 1$, it is necessary that

$$\int_{-\infty}^{\infty} f(x)dx = 1. \tag{17}$$

Equation (16) allows us to express the probability that X is in the interval (a, b) (for a continuous random variable it does not matter whether the end points are included or not), in terms of the probability density function:

$$P(a < X < b) = \int_{a}^{b} f(x)dx. \tag{18}$$

Example

5. Consider a random variable, X, which is equally likely to take any of its possible values. Such a random variable is said to be 'uniformly distributed' on its range. In the discrete case, if X can take n possible values, say $1, 2, \ldots, n$, then it is uniformly distributed on that range when

$$P(X = i) = p_i = 1/n; \quad i = 1, 2, \ldots, n.$$

In the continuous case, the range of values is usually an interval on the real line: $a < X < b$. Now, 'equally likely to take any possible value' means that the probability density function of X is a constant, c, on the interval (a, b), and is 0 elsewhere:

$$f(x) = \begin{cases} c & \text{if } a < x < b \\ 0 & \text{otherwise.} \end{cases}$$

On the other hand, in order that the normalising condition (17) be satisfied, the constant c must be equal to

$$c = \frac{1}{b-a}.$$

The distribution function of X is obtained from (16):

$$F(x) = \begin{cases} 0 & \text{if } x \leqslant a \\ (x-a)/(b-a) & \text{if } a < x < b \\ 1 & \text{if } b \leqslant x. \end{cases}$$

The probability density and distribution functions of a uniform random variable on an interval are shown in figure 1.2.

Figure 1.2

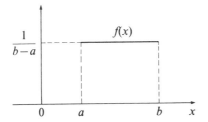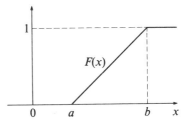

Of all the distribution functions on the interval (a, b), the uniform one offers least help in predicting the value that the random variable will take. If (s, t) is any sub-interval of (a, b), then the probability that X takes a value in that sub-interval depends only on its length and not on its position:

$$P(s < X < t) = F(t) - F(s) = \frac{t-s}{b-a}.$$

Suppose now that X and Y are two arbitrary random variables defined on the same sample space Ω. The two-variable function $F(x, y)$, defined by

$$F(x, y) = P(X < x, Y \leqslant y), \tag{19}$$

is called the 'joint distribution function' of X and Y. Knowledge of the joint distribution function implies that of the 'marginal' distribution functions of X and Y on their own. These are denoted by $F_X(x)$ and $F_Y(y)$ respectively. We have

$$F_X(x) = P(X \leqslant x) = P(X \leqslant x, Y \leqslant \infty) = F(x, \infty) \tag{20}$$

(the second equality uses the fact that the intersection of the events $X \leqslant x$

and $Y \leqslant \infty$ is equal to the event $X \leqslant x$). Similarly,

$$F_Y(y) = P(Y \leqslant y) = F(\infty, y). \tag{21}$$

In addition to characterising X and Y separately, the joint distribution function characterises their interaction. The two random variables are said to be 'independent of each other' if $F(x, y)$ factorises into a product of the two marginal distribution functions, i.e. if

$$F(x, y) = F(x, \infty)F(\infty, y) = F_X(x)F_Y(y). \tag{22}$$

In other words, in order that the random variables X and Y be independent, the events $\{X \leqslant x\}$ and $\{Y \leqslant y\}$ must be independent, for every x and y.

When X and Y are continuous, their joint probability density function, $f(x, y)$ is defined as

$$f(x, y) = \frac{\partial^2 F(x, y)}{\partial x \partial y}, \tag{23}$$

at the points where the partial derivative exists. As in the case of a single random variable, we can interpret this by writing

$$f(x, y) = \lim_{\substack{\Delta x \to 0 \\ \Delta y \to 0}} \frac{P(x < X \leqslant x + \Delta x, y < Y \leqslant y + \Delta y)}{\Delta x \Delta y}. \tag{24}$$

As a shorthand for (24), we may sometimes write

$$P(X = x, Y = y) = f(x, y)\mathrm{d}x\mathrm{d}y, \tag{25}$$

where $\mathrm{d}x$ and $\mathrm{d}y$ are infinitesimal quantities.

Given a joint probability density function, $f(x, y)$, the corresponding joint distribution function is obtained from

$$F(x, y) = \int_{-\infty}^{x} \int_{-\infty}^{y} f(u, v)\mathrm{d}u\mathrm{d}v. \tag{26}$$

In general, if S is a region in the two-dimensional (x, y) plane, the probability that the random point (X, Y) falls inside S is given by

$$P((X, Y) \in S) = \iint_S f(x, y)\mathrm{d}x\mathrm{d}y. \tag{27}$$

From the joint PDF, $f(x, y)$, one can derive the marginal PDFs of X and Y:

$$f_X(x) = \int_{-\infty}^{\infty} f(x, v)\mathrm{d}v; \quad f_Y(y) = \int_{-\infty}^{\infty} f(u, y)\mathrm{d}u. \tag{28}$$

The first of these equations is obtained by setting $y = \infty$ in (26) and then taking the derivative with respect to x; the second by setting $x = \infty$ and differentiating with respect to y.

It is also of interest to consider the 'conditional density function' of X given Y. This is defined as

$$f_{X|Y}(x|y) = \begin{cases} f(x, y)/f_Y(y) & \text{if } f_Y(y) > 0 \\ 0 & \text{if } f_Y(y) = 0. \end{cases} \tag{29}$$

Taken as a function of x, for a fixed y, the right-hand side of (29) is a probability density function for the random variable X, given that the random variable Y has taken value y. This can be justified by using the conditional probability formula (6), together with expressions (15) and (25):

$$P(X = x \mid Y = y) = \frac{P(X = x, Y = y)}{P(Y = y)} = \frac{f(x, y)\mathrm{d}x\mathrm{d}y}{f_Y(y)\mathrm{d}y} = \frac{f(x, y)}{f_Y(y)}\mathrm{d}x.$$

The condition for independence (22) can be stated in terms of probability density functions: two (continuous) random variables X and Y are mutually independent if

$$f(x, y) = f_X(x)f_Y(y). \tag{30}$$

As can be expected, if X and Y are independent, then the conditional PDF of X given Y is the same as the unconditional PDF of X:

$$f_{X|Y}(x|y) = f_X(x).$$

Example

6. Consider an experiment consisting of tossing two true dice. All 36 possible outcomes are equally likely. Let X, Y and Z be the random variables representing the numbers shown on the first die, the second die and the total on both dice, respectively. Then, since $P(X = i) = 1/6$, $P(Y = j) = 1/6$ and $P(X = i, Y = j) = 1/36$ $(i, j = 1, \ldots, 6)$, it is not difficult to see that the joint distribution function of X and Y factorises into a product of the marginal distribution functions. Hence, X and Y are independent random variables. However, X and Z are dependent, because $P(X \leqslant 1, Z \leqslant 2) = 1/36$, whereas $P(X \leqslant 1)P(Z \leqslant 2) = 1/216$.

Exercises

1. Let X be a random variable with probability distribution function $F(x)$, and let $x_1, x_2, \ldots$ be a monotone decreasing sequence converging to x. For $i = 1, 2, \ldots$, let A_i be the events defined by $A_1 = \{X > x_1\}$;

$A_i = \{x_i < X \leqslant x_{i-1}\}$ $(i > 1)$. Show that

$$P(X > x_n) = \sum_{i=1}^{n} P(A_i),$$

and

$$P(X > x) = \sum_{i=1}^{\infty} P(A_i).$$

Hence prove that $F(x) = \lim_{n \to \infty} F(x_n)$.

2. The distribution function of the 'rainfall' random variable, X, from example 2, is given by

$$F(x) = \begin{cases} 0 & \text{if } x < 0 \\ 0.01x^2 & \text{if } 0 \leqslant x \leqslant 10 \\ 1 & \text{if } x > 10. \end{cases}$$

Find the probability that the amount of rainfall is less than 1 or greater than 9. If an 'adjusted rainfall' random variable, Y, is defined by

$$Y = \begin{cases} X & \text{if } X \leqslant 6 \\ X - 1 & \text{if } X > 6, \end{cases}$$

find the distribution function of Y. Is Y continuous?

3. Let X and Y be two random variables with joint probability distribution function $F(x, y)$, and let S be the rectangle $S = \{(x, y); a < x \leqslant b, c < y \leqslant d\}$. Show that the probability that the random point (X, Y) falls inside S is given by

$$P((X, Y) \in S) = F(b, d) - F(a, d) - F(b, c) + F(a, c).$$

(Hint: find first the probabilities $P(a < X \leqslant b, Y \leqslant d)$ and $P(a < X \leqslant b, Y \leqslant c)$.)

4. Let X and Y be two random variables with joint PDF $f(x, y)$. Using the definition of conditional probability density function, obtain an expression for $f_{Y|X}(y|x)$ in terms of $f_Y(y)$ and $f_{X|Y}(x|y)$ only (that expression is the continuous analogue of the Bayes formula (10)).

1.3 Expectation and other moments

A random variable is characterised completely by its probability distribution function (or by its probability density function). However, such a full characterisation is not always easy to obtain, nor is it always absolutely necessary. It is sometimes desirable, and sufficient for practical purposes, to describe a random variable by one, or several, numbers which somehow summarise its essential attributes. Descriptors of this type exist.

They are defined in terms of the distribution function, but can often be determined directly, without having to know that function.

One of the most important characteristics of a random variable, X, is its 'mean', or 'average', or 'expectation', denoted by $E(X)$. This is defined as the (Riemann–Stieltjes) integral

$$E(X) = \int_{-\infty}^{\infty} x\,dF(x), \tag{31}$$

where $F(x)$ is the probability distribution function of X.

One can give a physical interpretation to the above quantity, and at the same time provide an intuitive justification for calling it a 'mean'. Let us replace, temporarily, the word 'probability' with the word 'mass', and think of $F(x)$ as describing the distribution of one unit of mass along the real line. That is, the mass is spread in such a way that $F(x)$ of it is to the left of, or at, point x, for every real x. Then the centre of gravity of the resulting line is precisely at the point given by the right-hand side of (31).

If X is a discrete random variable, taking values $x_1, x_2, \ldots$ with probabilities $p_1, p_2, \ldots$ respectively (in the physical analogy, the mass is concentrated at points $x_1, x_2, \ldots$, in amounts $p_1, p_2, \ldots$), then the integral in (31) becomes a sum:

$$E(X) = \sum_i x_i p_i. \tag{32}$$

This expression is easier to see as an average: a weighted mean of all the values that the random variable can take, where the weight of each value is the probability with which it is taken.

If X is a continuous random variable with PDF $f(x)$, then (31) can be rewritten as

$$E(X) = \int_{-\infty}^{\infty} xf(x)\,dx. \tag{33}$$

This is the continuous analogue of (32), since, according to (15), X takes value x with probability $f(x)\,dx$.

If X is neither discrete nor continuous, i.e. if $F(x)$ has continuously changing portions as well as jumps, then the general integral (31) has to be employed. Its evaluation in most practical cases reduces to that of terms like

$$\int_{a_i}^{b_i} xf(x)\,dx$$

over the continuous portions, plus terms of type

$$x_i[F(x_i) - F(x_i^-)]$$

at points where $F(x)$ has jumps.

Implicit in the definition of expectation is an assumption that the integral in the right-hand side of (31) converges. If it does not, then we say that $E(X)$ does not exist.

Examples

1. A gambler pays £3 for the privilege of throwing a single die. If the number that comes up is greater than 3, he will win that number of pounds; otherwise he will get nothing. Let X be the difference between the money gained and spent. This is a random variable whose value is -3 with probability $1/2$ (if 1, 2 or 3 comes up), 1 with probability $1/6$ (if 4 comes up), 2 with probability $1/6$ (if 5 comes up) and 3 with probability $1/6$ (if 6 comes up). We are assuming that the six possible outcomes are equally likely. Hence, the expected value of X is

$$E(X) = -\frac{3}{2} + \frac{1}{6} + \frac{2}{6} + \frac{3}{6} = -\frac{1}{2},$$

i.e. the gambler will lose 50 pence on the average.

2. A random variable, X, takes value i with probability $1/2^i$ $(i = 1, 2, \ldots)$. For instance, X could be the number of task execution attempts in example 3 of section 1.1. The expectation of X is obtained as

$$E(X) = \sum_{i=1}^{\infty} \frac{i}{2^i} = 2.$$

3. The probability density function of the rainfall random variable X (see exercise 2 of section 1.2) is given by

$$f(x) = \begin{cases} 0.02x & \text{if } 0 \leqslant x \leqslant 10 \\ 0 & \text{otherwise.} \end{cases}$$

Expression (33) yields the average amount of rainfall:

$$E(X) = 0.02 \int_0^{10} x^2 dx = 6\frac{2}{3}.$$

4. Let X be a random variable whose PDF is

$$f(x) = \frac{1}{\pi(1 + x^2)}; \quad -\infty < x < \infty.$$

This is known as the 'Cauchy density'. The variable X can be interpreted as the tangent of an angle (in radians) drawn at random from the range $(-\pi/2, \pi/2)$. This random variable has no expectation, since the integral

$$\int_{-\infty}^{\infty} xf(x)\,dx = \frac{1}{\pi} \int_{-\infty}^{\infty} \frac{x\,dx}{1+x^2}$$

does not converge.

The notions of probability and expectation are, in a certain sense, equally fundamental. To show that, consider an arbitrary event, A, in the sample space Ω. Define a random variable $I_A(\omega)$, for $\omega \in \Omega$, as follows:

$$I_A(\omega) = \begin{cases} 1 & \text{if } \omega \in A \\ 0 & \text{otherwise.} \end{cases}$$

In other words, I_A takes value 1 if A occurs and 0 if it does not. That random variable is called the 'indicator' of the event A. An application of (32) shows immediately that the expectation of I_A is equal to the probability of A:

$$E(I_A) = 1 \cdot P(I_A = 1) + 0 \cdot P(I_A = 0) = P(I_A = 1) = P(A). \tag{34}$$

Thus, instead of starting with an axiomatic treatment of probability and then defining expectation as in (31), one could take expectation as a starting point and define probability by means of (34).

The following properties of expectation are simple consequences of the definition (31). These properties would be the defining axioms if the alternative approach was adopted.

1. If $X \geqslant 0$ then $E(X) \geqslant 0$. Indeed, if X takes only non-negative values, then the integral in (31) can be taken over the semi-axis $0 \leqslant x < \infty$ and is therefore non-negative.

2. If c is a constant, then $E(cX) = cE(X)$. The random variable cX takes the value cx with the same probability as that with which X takes the value x. Hence,

$$E(cX) = \int_{-\infty}^{\infty} cx\,dF(x) = cE(X).$$

3. If X and Y are two random variables, then $E(X+Y) = E(X) + E(Y)$. We shall demonstrate this property for the case when X and Y are continuous, with joint PDF $f(x,y)$. The general case is treated similarly. Bearing in mind that when X takes the value x and Y takes the value y (which happens with probability $f(x,y)\,dx\,dy$), $X+Y$ takes the value $x+y$, we obtain

$$E(X + Y) = \int_{-\infty}^{\infty} \int_{-\infty}^{\infty} (x + y)f(x, y)\mathrm{d}x\,\mathrm{d}y$$

$$= \int_{-\infty}^{\infty} x \left[\int_{-\infty}^{\infty} f(x, y)\mathrm{d}y \right] \mathrm{d}x + \int_{-\infty}^{\infty} y \left[\int_{-\infty}^{\infty} f(x, y)\mathrm{d}x \right] \mathrm{d}y$$

$$= \int_{-\infty}^{\infty} x f_X(x)\mathrm{d}x + \int_{-\infty}^{\infty} y f_Y(y)\mathrm{d}y = E(X) + E(Y),$$

where we have used (28). Note that this property holds regardless of whether X and Y are independent or not.

4. $E(1) = 1$. This follows immediately from (32), which now has a single term.

5. If $X_1, X_2, \ldots$ is a monotonically increasing (decreasing) sequence of random variables, converging to X, then

$$E(X) = \lim_{n \to \infty} E(X_n).$$

This is a consequence of the fact that limit and integration can be interchanged.

From property 1 it follows that inequalities between random variables are preserved after taking expectations. In other words,

$$\text{if} \quad X \leqslant Y \quad \text{then} \quad E(X) \leqslant E(Y). \tag{35}$$

The reverse implication does not necessarily hold.

Properties 2 and 3 imply that if $X_1, X_2, \ldots, X_n$ are arbitrary random variables and $c_1, c_2, \ldots, c_n$ are constants, then

$$E \left[\sum_{i=1}^{n} c_i X_i \right] = \sum_{i=1}^{n} c_i E(X_i). \tag{36}$$

The expectation is unique among other possible quantities attempting to capture the essence of 'central value' (e.g. the 'median', m, defined by the equation $F(m) = 1/2$), in that it alone has these properties of a linear operator.

The next most important characteristic of a random variable, X, is its 'variance', denoted by $\mathrm{Var}(X)$. This is a measure of the 'spread' of the random variable around its mean. More precisely, it is defined as the average value of the square of the distance between X and its average value:

$$\mathrm{Var}(X) = E([X - E(X)]^2). \tag{37}$$

Clearly, if X takes values close to the mean $E(X)$ with high probability, then the variance $\mathrm{Var}(X)$ is small. Much less obvious, but also true, is the converse assertion: if $\mathrm{Var}(X)$ is small, then the probability that X takes a value close to $E(X)$ is large. Indeed, for any positive real number d, the probability that the distance between X and $E(X)$ is less than d is bounded from below by $1 - \text{Var}(X)/d^2$:

$$P(|X - E(X)| < d) \geqslant 1 - \frac{\text{Var}(X)}{d^2}. \tag{38}$$

This result is known as the 'Chebichev's inequality'. To prove it, consider the event $A = \{|X - E(X)| < d\}$, and let I_A be its indicator. The following inequality is satisfied whatever the outcome of the experiment:

$$I_A \geqslant 1 - \frac{[X - E(X)]^2}{d^2}. \tag{39}$$

(if A occurs, then I_A is 1 and the right-hand side is less than 1; if A does not occur, then I_A is 0 and the right-hand side is non-positive). Taking expectations in (39), and remembering that $E(I_A) = P(A)$, yields (38).

A trivial corollary of Chebichev's inequality is that, if $\text{Var}(X)$ is 0, then X is equal to its mean with probability 1, i.e. X is almost certainly a constant.

Let us now return to the expression for the variance (37) and rewrite it in a simpler form by opening the square brackets in the right-hand side:

$$\begin{aligned} \text{Var}(X) &= E(X^2 - 2XE(X) + [E(X)]^2) \\ &= E(X^2) - 2E(X)E(X) + [E(X)]^2 \cdot \\ &= E(X^2) - [E(X)]^2. \end{aligned} \tag{40}$$

The quantity $E(X^2)$ is of interest in its own right; it is called the 'second moment' of the random variable X. More generally, the expectation of the nth power of the random variable is called its 'nth moment' and is denoted by M_n:

$$M_n = E(X^n); \quad n = 1, 2, \ldots. \tag{41}$$

It can be shown that, under certain quite general conditions, a random variable is completely characterised by the set of all its moments. In other words, if $E(X^n)$ is known for all $n = 1, 2, \ldots$, then the distribution function of X is determined uniquely.

Using (40), we can write an expression for the variance of the sum of two random variables, X and Y:

$$\begin{aligned} \text{Var}(X + Y) &= E[(X + Y)^2] - [E(X) + E(Y)]^2 \\ &= E(X^2) + 2E(XY) + E(Y^2) - [E(X)]^2 \\ &\quad - 2E(X)E(Y) - [E(Y)]^2 \\ &= \text{Var}(X) + \text{Var}(Y) + 2[E(XY) - E(X)E(Y)]. \tag{42} \end{aligned}$$

Clearly, the variance of a sum is only equal to the sum of variances if the following quantity vanishes:

$$\text{Cov}(X,Y) = E(XY) - E(X)E(Y). \tag{43}$$

That quantity is called the 'covariance' of X and Y. It is a measure of the extent to which X and Y are correlated. If $\text{Cov}(X,Y) = 0$, then X and Y are said to be 'uncorrelated'. It is not difficult to see that, if X and Y are independent, then they are uncorrelated. However, it is possible that X and Y are uncorrelated, and yet dependent.

Exercises

1. Find the mean and the variance of the uniformly distributed random variables defined in example 5 of section 1.2.

2. Let X be a non-negative random variable and d a positive constant. Consider the event $A = \{X > d\}$ and show that its indicator satisfies the inequality $I_A \leqslant X/d$. Hence derive the 'Markov inequality'

$$P(X > d) \leqslant \frac{E(X)}{d}.$$

3. Show that, for any two random variables X and Y,

$$[E(XY)]^2 \leqslant E(X^2)E(Y^2).$$

This is known as the Cauchy–Schwarz inequality. (Hint: note that

$$E(X^2) - 2aE(XY) + a^2 E(Y^2) = E[(X - aY)^2]$$

is non-negative for every a. Hence, the discriminant of the quadratic form in the left-hand side must be less than or equal to 0.)

4. The 'correlation coefficient' of X and Y, $r_{X,Y}$, is defined as

$$r_{X,Y} = \frac{\text{Cov}(X,Y)}{[\text{Var}(X)\,\text{Var}(Y)]^{1/2}}.$$

Show that that coefficient is always in the range $-1 \leqslant r_{X,Y} \leqslant 1$. (Hint: apply the Cauchy–Schwarz inequality to the random variables $X - E(X)$ and $Y - E(Y)$.)

5. Let X and Y be two random variables and a and b be two constants. Show that

$$\text{Var}(aX + bY) = a^2\,\text{Var}(X) + 2ab\,\text{Cov}(X,Y) + b^2\,\text{Var}(Y).$$

Generalise this expression to a linear combination of n random variables.

6. Show that, if the expectation $E(X)$ exists, it can be obtained from

$$E(X) = \int_0^\infty [1 - F(x)]dx - \int_{-\infty}^0 F(x)dx.$$

(Hint: integrate the above expression by parts and reduce it to (31).)

1.4   Bernoulli trials and related random variables.

We shall examine here what is perhaps the simplest non-trivial example of a random process, i.e. an experiment which involves time and where different things may happen at different moments. Imagine first an action which may have two possible outcomes, such as tossing a coin, or exposing the top card of a well-shuffled pack in order to see whether it is an ace or not. We shall call such an action a 'trial' and its two outcomes 'success' and 'failure'.

Now consider an experiment which consists of performing a sequence of identical trials, so that the outcome of trial i is independent of the outcomes of trials $1, 2, \ldots, i - 1$. Moreover, suppose that the outcome of each trial is a success with probability q and a failure with probability $1 - q$ ($0 < q < 1$). Such an experiment is called 'Bernoulli trials'; its outcomes are infinite sequences of successes and failures.

A random variable of interest in a Bernoulli trials experiment is the index, K, of the trial at which the first success occurs. Because of the nature of the experiment, the number of trials between the first and the second successes (excluding the former but including the latter), or between any two consecutive successes, has the same distribution as K.

To find the probability, p_k, that K takes the value k, note that the event $\{K = k\}$ occurs if, and only if, the first $k - 1$ trials are failures and the kth one is a success. Hence,

$$p_k = P(K = k) = (1 - q)^{k-1}q; \quad k = 1, 2, \ldots. \tag{44}$$

From this expression, the distribution function of K is obtained as

$$F(k) = P(K \leqslant k) = \sum_{i=1}^k p_i = 1 - (1 - q)^k; \quad k = 1, 2, \ldots. \tag{45}$$

This is known as the 'geometric distribution'. That it is, indeed, a distribution, follows from the fact that

$$F(\infty) = \sum_{k=1}^\infty p_k = 1.$$

In other words, a success will occur after a finite number of trials with probability 1. The experiment may, of course, result in all trials being failures. However, the probability of such an outcome is 0.

The average value of K is given by

$$E(K) = \sum_{k=1}^{\infty} kp_k = q \sum_{k=1}^{\infty} k(1-q)^{k-1} = \frac{1}{q}. \tag{46}$$

As can be expected, the lower the probability of success, the longer one has to wait, on the average, until the first success.

The geometric distribution has a property which makes it unique among discrete distributions. Formally, this can be stated as follows: if i and j are any two positive integers, then

$$1 - F(i+j) = [1 - F(i)][1 - F(j)]. \tag{47}$$

This property, which is an immediate consequence of (45), also has an obvious physical interpretation: if S_1 is a set of i consecutive trials and S_2 is the set of the following j consecutive trials, then the events 'no successes occur during S_1' and 'no successes occur during S_2' are independent of each other. That is precisely what (47) says:

$$P(K > i+j) = P(K > i)P(K > j); \quad i, j \geqslant 1. \tag{48}$$

Conversely, if (47) holds for some discrete distribution, $F(k)$, $k = 1, 2, \ldots$, then that distribution must have the form (45), i.e. it must be geometric (see exercise 1).

From (48) it follows that the conditional probability that there are no successes during S_1 and S_2, given that there are no successes during S_1, is equal to the unconditional probability that there are no successes during S_2:

$$P(K > i+j \,|\, K > i) = \frac{P(K > i+j)}{P(K > i)} = P(K > j). \tag{49}$$

This is sometimes referred to as the 'memoryless property' of the geometric distribution.

Examples

1. A processor has an infinite supply of jobs, which it executes one after the other. The execution of a job consists of giving it one or more quanta of service, each lasting exactly one time unit. The job is completed at the end of a service quantum with probability q, and requires more service with probability $1 - q$ ($0 < q < 1$); each such outcome is independent of all previous ones.

The consecutive service quanta can thus be considered as Bernoulli trials, a job completion being a success. Therefore, the number of service quanta required by a job, and hence the job execution time, is distributed geometrically with parameter q. According to the memoryless property, the probability that a job will require at least j more time units to complete, given that it has already been running for i time units, is the same as the unconditional probability that it needs at least j time units.

2. Consider a 'slotted' communication channel which can accept packets of information for transmission at time instants $0, 1, 2, \ldots$. A packet takes exactly one time unit to transmit. At time i, there is either one packet available for transmission, with probability q, or no packets, with probability $1 - q$, regardless of what happened at previous instants.

The channel goes through alternating periods of being idle (having nothing to transmit), and busy (transmitting a sequence of packets). Suppose that the first slot is an idle one, i.e. there was no packet at time 0. Then, considering the instants $1, 2, \ldots$ as Bernoulli trials, success occurring when a packet is present, we see that the length of the first idle period is determined by the index of the first success. That length, and hence the length of any other idle period, is therefore distributed geometrically with parameter q. Similarly, supposing that the first slot is busy, and treating the absence of a packet as a success, we conclude that the duration of any busy period is distributed geometrically with parameter $1 - q$.

3. In the above example, let us change the definition of busy and idle periods by postulating that every busy slot is a separate busy period (of length 1). Between two consecutive busy periods, there is an idle period which may be of length 0. Now, if J is the length of an idle period, J takes value j when j idle slots are followed by a busy slot ($j = 0, 1, \ldots$). Hence,

$$P(J = j) = (1 - q)^j q; \quad j = 0, 1, \ldots$$

and

$$P(J \leqslant j) = 1 - (1 - q)^{j+1}; \quad j = 0, 1, \ldots.$$

We shall sometimes refer to this as the 'modified geometric' distribution. The average value of J is given by

$$E(J) = \sum_{j=0}^{\infty} j(1 - q)^j q = \frac{1 - q}{q}.$$

Another random variable of interest in connection with a Bernoulli trials experiment is the number of successes, S, that occur during the first n trials (or during any n trials). If we select a particular set of s trials ($s \leqslant n$), then the probability that those trials are successes and the other $n - s$ trials are failures is equal to $q^s(1 - q)^{n-s}$. On the other hand, the number of ways of selecting s out of n possibilities is equal to the binomial coefficient

$$\binom{n}{s} = \frac{n!}{s!\,(n-s)!} \, .$$

Hence, the probability that there are s successes among n trials is given by

$$P(S = s) = \binom{n}{s} q^s (1 - q)^{n-s}; \quad s = 0, 1, \ldots, n. \tag{50}$$

This is known as the 'binomial distribution'. The random variable S can also be expressed as

$$S = \sum_{i=1}^{n} X_i, \tag{51}$$

where X_i is the number of successes that occur on the ith trial. Obviously, X_i can only take the values 0 and 1, and does not depend on i. The mean and variance of X_i are easily derived:

$$E(X_i) = 1 \cdot P(X_i = 1) = q;$$
$$E(X_i^2) = 1 \cdot P(X_i^2 = 1) = q;$$
$$\mathrm{Var}(X_i) = E(X_i^2) - [E(X_i)]^2 = q(1 - q). \tag{52}$$

Taking expectations in (51) we obtain

$$E(S) = nq. \tag{53}$$

Similarly, since the random variables X_i are mutually independent, the variance of their sum is equal to the sum of their variances:

$$\mathrm{Var}(S) = nq(1 - q). \tag{54}$$

Expressions (53) and (54) can also be derived directly from (50).

Consider the random variable S/n, which represents the frequency with which successes occur within the sequence of n trials. The average value of that random variable is, according to (53), equal to q for every n. This by itself does not, of course, imply anything about the value that S/n will take in any particular realisation of the experiment. However, it is an empirical fact that, if n is large, the fraction of successes, S/n, is likely to be close to the probability of success, q. Indeed, before the axiomatic treatment of probability was developed, the natural scientists used to define the probability of an event as the fraction of the experiments in which the event occurred. We can now supply a mathematical confirmation of this empirical fact. It can be shown that, whatever the positive number ε, the probability that S/n will deviate from q by less than ε approaches 1 as n increases:

$$\lim_{n \to \infty} P\left(\left|\frac{S}{n} - q\right| < \varepsilon\right) = 1. \tag{55}$$

The proof of this assertion, known as the 'weak law of large numbers', is outlined in exercise 3. Note that, in view of (51), S/n is the 'arithmetic mean', or the 'sample mean', of the random variables $X_1, X_2, \ldots, X_n$. If those were arbitrary independent random variables (rather than being indicators of success), with mean q and finite variance, (55) would continue to hold. In fact, an even stronger statement can be made. We can 'almost' guarantee that the sample mean S/n will converge to the expectation q in any given realisation of the experiment:

$$P\left(\lim_{n \to \infty} \frac{S}{n} = q\right) = 1. \tag{56}$$

In other words, although there may be experimental outcomes for which S/n does not converge to q, the probability of such outcomes is 0. This is known as the 'strong law of large numbers'.

In example 1 above, the number of jobs completed by the server during n service quanta has the binomial distribution. The expectation of that number is nq and hence the average system throughput is q jobs per unit time. Similarly, in example 2, the number of packets transmitted during n slots is binomially distributed and the average channel throughput is q packets/unit time.

There are many applications of the binomial distribution in the areas of quality control and reliability. If each one of a series of items has a probability q of being defective, independently of the others, and a sample of n items is given, then the number, S, of defective items in that sample is binomially distributed. Moreover, according to the laws of large numbers, the fraction S/n can be taken as an estimate of q. In this context, 'item' may mean a manufactured object, a hardware component or a software module.

Exercises

1. Let $a_1, a_2, \ldots$ be a sequence of numbers in the interval $(0, 1)$, which satisfy the equations $a_{i+j} = a_i a_j$, for $i, j = 1, 2, \ldots$. Show, by setting $j = 1$ and solving the resulting recurrence equations, that $a_i = a_1^i$, $i = 1, 2, \ldots$. Hence demonstrate that any distribution which satisfies (48) must be of the form (45).

2. Use (44) to find the second moment of the geometrically distributed random variable K. Then show that the variance of K is given by

$$\mathrm{Var}(K) = q^{-2}(1 - q).$$

3. Establish (55) by applying Chebichev's inequality to the random variable S/n. (Hint: the variance of S/n approaches 0 as n increases.)

4. In a Bernoulli trials experiment, let N_s be the index of the trial at which the sth success occurs, i.e. the number of trials required to achieve s successes. Bearing in mind that, in order for N_s to take value n, the nth trial must be a success and there must be exactly $s - 1$ successes among the preceding $n - 1$ trials, show that the distribution of N_s, known as the 'negative binomial distribution', is given by

$$P(N_s = n) = \binom{n - 1}{s - 1} q^s (1 - q)^{n-s}; \quad n = s, s + 1, \ldots .$$

From here, or from the fact that N_s can be thought of as the sum of s independent geometrically distributed random variables, derive the mean and variance of N_s as

$$E(N_s) = \frac{s}{q}; \quad \mathrm{Var}(N_s) = sq^{-2}(1 - q).$$

5. In order to assess how well a student has mastered a new programming language, he is asked to write and run five unrelated programs. He will be considered to have passed the test if at least three of them are correct. Assuming that each program the student writes is correct with probability 0.6 (independently of the others), what is the probability that he will pass?

1.5   Sums, transforms and limits

We have already encountered instances where the object of interest is a sum of independent random variables. The mean and the variance of such a sum can, of course, be obtained by adding together, respectively, the means and the variances of the constituent variables (the former can be done without the assumption of independence). However, finding the distribution of the sum is, in general, a more difficult task which may involve a considerable computational effort.

Consider first the sum, S, of two independent discrete random variables, X and Y, whose values are non-negative integers. Denote by p_i, q_i and r_i $(i = 0, 1, \ldots)$ the distributions of X, Y and S, respectively:

$$p_i = P(X = i); \quad q_i = P(Y = i); \quad r_i = P(S = i); \quad i = 0, 1, \ldots .$$

Since, in order for S to take the value i, X must take one of the values $0, 1, \ldots, i$ and Y must take the value $i - X$, we can write

$$r_i = P(X + Y = i) = \sum_{j=0}^{i} P(X = j, Y = i - j)$$

$$= \sum_{j=0}^{i} P(X = j)P(Y = i - j) = \sum_{j=0}^{i} p_j q_{i-j}; \quad i = 0, 1, \ldots. \quad (57)$$

The distribution r_i, obtained according to (57), is called the 'convolution' of the distributions p_i and q_i. Note that the reason for requiring X and Y to be independent is to ensure that the third equality in (57) holds. If X and Y are not independent, the distribution of their sum is not necessarily equal to the convolution of their distributions.

The convolution operation can be presented in another form, which is often much more convenient. Let us associate with the distribution p_i, the function $p(z)$, defined as

$$p(z) = \sum_{i=0}^{\infty} p_i z^i. \quad (58)$$

This is called the 'generating function', or the 'z-transform', of the probabilities p_i. That function satisfies the relation $p(1) = 1$, since all the p_is must sum up to 1. Therefore, $p(z)$ is finite at least for real values z in the interval $-1 \leqslant z \leqslant 1$. The term 'generating function' is justified by the fact that if $p(z)$ is known, then the probabilities p_i are determined uniquely:

$$p_i = \frac{1}{i!} p^{(i)}(0); \quad i = 0, 1, \ldots, \quad (59)$$

where $p^{(i)}(0)$ is the ith derivative of $p(z)$ at $z = 0$.

The generating function can also be expressed in the form of an expectation:

$$p(z) = E(z^X). \quad (60)$$

This follows directly from the definition of expectation, (32). It suffices to note that the random variable z^X takes the value z^i with probability p_i $(i = 0, 1, \ldots)$.

Now introduce also the generating functions, $q(z)$ and $r(z)$, of the distributions q_i and r_i, respectively. These are defined by power series of the type (58), and can be expressed in the form (60). We can then write

$$r(z) = E(z^S) = E(z^{X+Y}) = E(z^X z^Y)$$

$$= E(z^X)E(z^Y) = p(z)q(z). \quad (61)$$

Thus, the generating function of the convolution of two discrete distributions is equal to the product of their generating functions. Moreover, (61) generalises in an obvious way to sums of any finite number of independent random variables.

There are other important applications for generating functions, besides computing convolutions. For instance, we shall encounter problems where a discrete distribution has to be determined by solving an infinite system of linear, or differential, equations. An elegant way of dealing with such a system is to reduce it to a single equation for an unknown generating function, and then solve the latter.

Given a generating function, it is quite easy to find the mean, and higher moments, of the corresponding random variable. Indeed, differentiating (60) with respect to z yields

$$p'(z) = E(Xz^{X-1}).$$ (62)

Hence, the expectation of X is obtained by setting $z = 1$ in (62):

$$p'(1) = E(X).$$ (63)

Similarly, the second derivative of $p(z)$ at $z = 1$ gives

$$p''(1) = E[X(X-1)] = E(X^2) - E(X),$$ (64)

which can be used to determine $E(X^2)$. In general,

$$p^{(n)}(1) = E[X(X-1)\cdots(X-n+1)].$$ (65)

The expectation in the right-hand side of (65) is sometimes referred to as the 'factorial moment' of X, of order n.

Examples

1. We saw that a single trial in a Bernoulli trials experiment is associated with a random variable, X, which takes value 0 with probability $1 - q$ and 1 with probability q $(0 < q < 1)$. The generating function of this distribution is $q(z) = 1 - q + qz$. Therefore, the sum, S, of n independent such variables has a generating function given by

$$r(z) = (1 - q + qz)^n.$$ (66)

This, according to (51), is the generating function of the binomial distribution. Indeed, if the right-hand side of (66) is expanded in powers of z, the coefficient of z^s turns out to be exactly the right-hand side of (50). Expressions (53) and (54) can be obtained by taking derivatives in (66) at $z = 1$.

2. The generating function of the geometric distribution, (44), is obtained as

$$p(z) = \sum_{i=1}^{\infty} q(1-q)^{i-1} z^i = \frac{qz}{1 - (1-q)z}.$$ (67)

From here one could derive the mean, (46): $p'(1) = 1/q$. It also follows that the generating function of the negative binomial distribution (see exercise 4 of section 1.4) must be equal to $[p(z)]^s$, since that distribution is the s-fold convolution of the geometric one.

Let us now return to the distribution of the sum of independent random variables, this time in the case when the latter are continuous and have probability density functions. Let X and Y be two such random variables, with PDFs $f(x)$ and $g(x)$ respectively. Assume for the moment that X and Y take non-negative values. Then we can write an expression analogous to (57) for the PDF, $h(x)$, of the sum $S = X + Y$:

$$h(x) = \int_0^x f(t)g(x - t)\mathrm{d}t. \tag{68}$$

This is justified by arguing that, in order for S to take the value x (which happens with probability $h(x)\mathrm{d}x$), X must take some value $t < x$ (probability $f(t)\mathrm{d}t$) and Y must be equal to $x - t$ (probability $g(x - t)\mathrm{d}x$). Summing over all possible values for t gives the integral in the right-hand side of (68). That integral is also called the 'convolution' of the two functions f and g.

Just as in the case of discrete distributions, manipulations with probability density functions are often simplified considerably by the introduction of transforms. When dealing with non-negative valued random variables, it is convenient to associate with a probability density function, $f(x)$, its 'Laplace transform', $f^*(s)$, defined as

$$f^*(s) = \int_0^\infty \mathrm{e}^{-sx}f(x)\mathrm{d}x. \tag{69}$$

The right-hand side of (69) is finite for all $s \geq 0$ (and in fact for all complex s whose real part is non-negative), and satisfies the normalising condition $f^*(0) = 1$. It can be demonstrated that the Laplace transform $f^*(s)$ determines the probability density function $f(x)$ uniquely. Indeed, there is an explicit formula expressing f in terms of f^*, which may be used for a numerical inversion:

$$f(x) = \frac{1}{2\pi} \int_{-\infty}^\infty \mathrm{e}^{-isx}f^*(-is)\mathrm{d}s, \tag{70}$$

where $i = \sqrt{-1}$ is the imaginary unit.

It is readily seen, from the definition (33), that the Laplace transform can be written in the form of an expectation:

$$f^*(s) = E(\mathrm{e}^{-sX}). \tag{71}$$

From this, it follows immediately that the Laplace transform of the sum of two independent random variables is equal to the product of their Laplace transforms:

$$h^*(s) = E(e^{-s(X+Y)}) = E(e^{-sX}e^{-sY})$$

$$= E(e^{-sX})E(e^{-sY}) = f^*(s)g^*(s). \tag{72}$$

One can also obtain this result directly from (68). Note that, although the independence of X and Y is sufficient to ensure the validity of (72), it is not necessary. There are examples of random variables for which the result holds, despite the fact that they are dependent on each other.

The derivatives of a Laplace transform at point $s = 0$ yield the moments of the corresponding random variable. For instance, differentiating (71) with respect to s we get

$$f^{*'}(s) = E(-Xe^{-sX}).$$

Hence, setting $s = 0$,

$$E(X) = -f^{*'}(0). \tag{73}$$

In general, if the nth moment of X exists, then it is given by

$$E(X^n) = (-1)^n f^{*(n)}(0); \quad n = 0, 1, \ldots . \tag{74}$$

It may not be convenient, or even possible, to use Laplace transforms for random variables which take both positive and negative values. The integral in the right-hand side of (69), when taken over the interval $(-\infty, \infty)$, may diverge. To get around this difficulty, another transform of a similar type has been introduced. This is called the 'characteristic function', $\varphi_X(s)$, of the random variable X, and is defined as follows:

$$\varphi_X(s) = E(e^{isX}); \quad -\infty < s < \infty. \tag{75}$$

Here, depending on whether X is discrete or continuous, the expectation in the right hand side is given by a sum or an integral. The advantage of this definition is that, since e^{isX} is bounded, $\varphi_X(s)$ exists for all X (see exercise 2). The disadvantage, such as it is, lies in having to deal with complex numbers.

Like the other transforms that we have seen, the characteristic function of a random variable determines its distribution uniquely. If X and Y are independent, then

$$\varphi_{X+Y}(s) = \varphi_X(s)\varphi_Y(s). \tag{76}$$

If the nth moment of X exists, then it can be obtained from

$$E(X^n) = i^{-n}\varphi_X^{(n)}(0); \quad n = 0, 1, \ldots . \tag{77}$$

Another direct consequence of the definition (75) is that, if Y is obtained from X by a linear transformation, i.e. $Y = aX + b$, then

$$\varphi_Y(s) = e^{isb}\varphi_X(as). \tag{78}$$

In the modelling applications that will concern us in this book, the random variables of interest will almost always be non-negatively valued. Therefore, when we need to use transform methods, we shall be able to restrict ourselves to generating functions and Laplace transforms. However, before we leave this section, it will be instructive to examine a random variable for which the use of the characteristic function is indicated. At the same time, we shall introduce a distribution that plays a very important role in probability theory.

Let X be a random variable whose probability density function is given by

$$f(x) = \frac{1}{\sqrt{2\pi}} e^{-x^2/2}; \quad -\infty < x < \infty. \tag{79}$$

This is called the 'standard normal' density. Its shape is illustrated in figure 1.3.

Figure 1.3

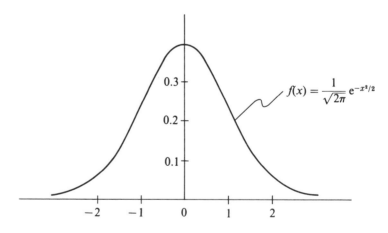

The distribution function corresponding to the density (79) is

$$F(x) = \frac{1}{\sqrt{2\pi}} \int_{-\infty}^{x} e^{-u^2/2} du. \tag{80}$$

The characteristic function of the standard normal random variable is equal to

$$\varphi(s) = \frac{1}{\sqrt{2\pi}} \int_{-\infty}^{\infty} e^{isx - x^2/2} dx = e^{-s^2/2}. \tag{81}$$

(The derivation of this result will be omitted: it requires contour integration and an application of Cauchy's theorem.)

The mean and variance of X are immediately obtained with the aid of (77):

$$E(X) = \frac{1}{i}\varphi'(0) = 0,$$

$$\mathrm{Var}(X) = E(X^2) = -\varphi''(0) = 1. \tag{82}$$

The standard normal distribution gives rise to a family of distributions depending on two parameters, μ and σ $(\sigma > 0)$. The random variable $Y = \mu + \sigma X$, which has mean μ and variance σ^2, is said to have the general normal distribution, denoted by $N(\mu, \sigma^2)$. Its characteristic function is, according to (78),

$$\varphi_Y(s) = \exp(i\mu s - \sigma^2 s^2/2). \tag{83}$$

It is easy to verify (see exercise 3) that the probability density function corresponding to (83) is

$$f_y(x) = \frac{1}{\sigma\sqrt{2\pi}} \exp\left[-\frac{1}{2}\left(\frac{x - \mu}{\sigma} \right)^2 \right]. \tag{84}$$

The first interesting property of the normal family is that it is closed with respect to convolution. In other words, if Y_1 and Y_2 are independent random variables with distributions $N(\mu_1, \sigma_1^2)$ and $N(\mu_2, \sigma_2^2)$ respectively, then their sum, $Y_1 + Y_2$, has the normal distribution $N(\mu_1 + \mu_2, \sigma_1^2 + \sigma_2^2)$. This is a direct consequence of (76) and (83): the product of two characteristic functions of type (83) is also of type (83).

It is equally easy to see that if Y has the distribution $N(\mu, \sigma^2)$ and c is an arbitrary real constant, then the random variable cY has the distribution $N(c\mu, c^2\sigma^2)$.

A simple corollary of the above is that if $X_1, X_2, \ldots, X_n$ are independent random variables, all having the standard normal distribution $N(0, 1)$, then the random variable $Z_n = (X_1 + X_2 + \cdots + X_n)/\sqrt{n}$ also has the standard normal distribution, no matter what the value of n. More remarkable, however, is the fact (which we state without proof) that $N(0, 1)$ is the only continuous distribution with this 'preservation' property.

Now consider a more general sum of independent and identically distributed random variables. Assume that $X_1, X_2, \ldots, X_n$ have mean 0 and variance 1, but otherwise allow their distribution to be arbitrary. Then it is possible to assert that the sum $Z_n = (X_1 + X_2 + \cdots + X_n)/\sqrt{n}$ has approximately the standard normal distribution. More precisely, the following result holds:

When $n \to \infty$, the distribution of Z_n approaches $N(0, 1)$.

This is a special case of what is known as the 'central limit theorem'. The use of characteristic functions reduces its proof to a few uncomplicated manipulations:

Let $\varphi(s)$ be the characteristic function of X_i. Then the characteristic function of Z_n, $\psi_n(s)$, is given by

$$\psi_n(s) = \left[\varphi\left(\frac{s}{\sqrt{n}} \right) \right]^n; \quad n = 1, 2, \ldots . \tag{85}$$

Now, since the first two moments of X_i exist and are equal to 0 and 1 respectively, a partial Taylor expansion for $\varphi(s)$ can be written, in the form

$$\varphi(s) = \varphi(0) + \varphi'(0)s + \tfrac{1}{2}\varphi''(0)s^2 + o(s^2)$$
$$= 1 - \tfrac{1}{2}s^2 + o(s^2), \tag{86}$$

where $o(x)$ is a function that tends to 0 faster than x, i.e. $o(x)/x \to 0$ when $x \to 0$.

Substituting (86) into (85) and letting $n \to \infty$, we get

$$\lim_{n \to \infty} \psi_n(s) = \lim_{n \to \infty} \left[1 - \frac{1}{2n}s^2 + o\left(\frac{s^2}{n} \right) \right]^n = e^{-s^2/2}. \tag{87}$$

This establishes the proposition, since the right-hand side of (87) is precisely the characteristic function of the standard normal distribution.

Having got this result, it is easy to deal with the case when the X_is have arbitrary (but finite) mean μ and variance σ^2. Then it suffices to note that the random variables $(X_i - \mu)/\sigma$ have mean 0 and variance 1. Hence, the normalised sum

$$\sum_{i=1}^{n} (X_i - n\mu)/(\sigma\sqrt{n})$$

approaches $N(0, 1)$, in distribution, when $n \to \infty$.

The central limit theorem is valid under considerably more general conditions than the ones stated here.

Exercises

1. Let S_n be the number of successes in a series of n Bernoulli trials, where the probability of success is q. Show that when $n \to \infty$ and $q \to 0$, in such a way that $nq = \lambda$ remains constant, the generating function of S_n (given by (66)) approaches $e^{-\lambda(1-z)}$. Hence deduce that, in the limit, the probability that there will be k successes is equal to

$$p_k = \frac{\lambda^k}{k!} e^{-\lambda}; \quad k = 0, 1, \ldots .$$

This is the 'Poisson distribution', which we shall encounter again later.

2. Using the fact that the expectation operator satisfies the inequality $|E(Y)| \leqslant E(|Y|)$, show that the characteristic function of an arbitrary random variable satisfies

$$|\varphi(s)| \leqslant 1,$$

with equality at $s = 0$.

3. Establish (83) by evaluating the integral

$$Y(s) = \int_{-\infty}^{\infty} e^{isx} f_Y(x) dx,$$

where $f_Y(x)$ is the general normal density, (84). Make a change of variables $u = (x - \mu)/\sigma$ and use (81).

4. Find the Laplace transforms of
   (i) the uniform density $f(x) = 1/(b-a)$, $a \leqslant x \leqslant b$;
   (ii) the 'exponential' density, $f(x) = \lambda e^{-\lambda x}$, $0 \leqslant x < \infty$ $(\lambda > 0)$.

Hence obtain the first two moments of those distributions.

Literature

There is no lack of good books on probability theory. For the reader wishing a solid background in the subject, we would recommend the excellent work by Feller [9]. The shorter, more recent and very readable text by Whittle [33] takes expectation as the fundamental concept, rather than probability. For the more mathematically inclined, a treatment of measure theory as a basis for probability can be found in Loeve [26] or Kingman and Taylor [19].

2
Stochastic processes

Computer systems, along with banks, railway stations and beehives, have two very fundamental properties.

(a) They are dynamic, i.e. they go through a succession of different states as time progresses. We shall refer to these successions of states as 'operation paths'.

(b) Their behaviour is influenced by random phenomena. That is, the observation of such a system over a period of time can be considered as a random experiment, whose possible outcomes are operation paths.

Suppose that we are interested in some numerical characteristic of the system, such as the number of tasks competing for service as a device, or the number of bees occupied in gathering nectar. Denote that characteristic, at a given time, t, by X_t. Clearly, the value taken by X_t depends on the operation path that the system is following. In other words, X_t is a random variable defined on the sample space, Ω, consisting of all possible operation paths during the observation period, T. To different points in T correspond different random variables, but the sample space is always Ω.

We are thus led to the realisation that the appropriate tool for modelling and studying the dynamic behaviour of systems is a parameterised family of random variables, $\{X_t; t \in T\}$, defined on the same sample space. Such a family is called a 'stochastic process'. The set T is the 'parameter space' of the process (we shall always think of the parameter t as representing time, although it does not have to). The set of values that X_t may take is the 'state space' of the process. Depending on the nature of those two sets, we may be dealing with a 'discrete parameter' or 'continuous parameter' process, and also with a 'discrete state' or 'continuous state' process.

Examples

1. Let X_t be the amount of water in a reservoir at time t after the latter is put into operation. Then $\{X_t; t \geq 0\}$ is a continuous parameter, continuous state stochastic process (the state space is an interval of the type $[0, C]$, where C is the capacity of the reservoir).

2. A customer receives monthly statements from his bank, describing the transactions performed. Let B_n be the balance of the account at the end of the nth month and C_n be the bank's charges for that month. Assuming that all amounts are given to the nearest penny, both $\{B_n; n = 1, 2, \ldots\}$ and $\{C_n; n = 1, 2, \ldots\}$ are discrete parameter, discrete state stochastic processes.

3. In a multiprocessor computer system, the number of busy processors at time t, K_t, the number of jobs present at time t, N_t, and the number of jobs completed up to time t, D_t, define the stochastic processes $\{K_t; t \geq 0\}$, $\{N_t; t \geq 0\}$ and $\{D_t; t \geq 0\}$. All three of these have a discrete state space and a continuous parameter.

2.1 Renewal processes

Consider a phenomenon which manifests itself first at time 0 and thereafter keeps occurring, at random intervals, *ad infinitum*. Denote the consecutive instants of occurrence by T_n ($n = 0, 1, \ldots$; $T_0 = 0$), and let $S_n = T_n - T_{n-1}$ ($n = 1, 2, \ldots$) be the times between them. Suppose further that the random variables S_n are independent and identically distributed.

The sequence $\{T_n; n = 0, 1, \ldots\}$ is called a 'renewal process'. The instants T_n are referred to as the 'renewal points'; the times, S_n, between consecutive renewal points are the 'renewal periods'. The name 'renewal' is justified by the fact that if the time origin is shifted to point T_n, for some $n \geq 1$, the subsequent process will be indistinguishable from the original.

The initial applications of renewal processes were in the field of machine reliability. A new machine starts work at time 0. After a while it breaks down and is replaced (either immediately or after some delay) by an identical new machine; and so on. The renewal points are then the replacement instants, while a renewal period consists of either the lifetime of a machine, or that lifetime plus the following replacement delay.

We shall use renewal processes mainly to model streams of jobs, or communication requests, arriving into, or departing from, a system. These applications will also affect the terminology. For instance, we shall often refer to the renewal points and renewal periods as 'arrival instants' and 'interarrival times', respectively.

A renewal process is completely characterised by specifying the common distribution function, $F(x)$, or the density function, $f(x)$ (if the latter exists), of the renewal periods. The nth renewal point can be expressed as a sum of n

renewal periods:

$$T_n = \sum_{i=1}^{n} S_n; \quad n = 1, 2, \ldots . \tag{1}$$

Hence, the distribution function of T_n, $F_n(x)$, can be obtained by taking the n-fold convolution of $F(x)$ with itself (see section 1.5). That is, in general, a difficult operation. However, for large values of n, the central limit theorem tells us that T_n is approximately normally distributed. More precisely, if the mean and variance of the renewal periods are m and v^2 respectively, then T_n has approximately the distribution $N(nm, nv^2)$. Alternatively, the random variable $(T_n - nm)/(v\sqrt{n})$ has approximately the standard normal distribution, $N(0, 1)$.

Consider now the number, K_t, of renewal points that fall in the interval $(0, t]$. Clearly, for that random variable to take value k, the kth renewal point must occur before, or at time t, and the $(k + 1)$st one must occur after t. Therefore, the distribution of K_t is given by

$$\begin{aligned} P(K_t = k) &= P(T_k \leqslant t < T_{k+1}) = P(t < T_{k+1}) - P(t < T_k) \\ &= 1 - F_{k+1}(t) - [1 - F_k(t)] = F_k(t) - F_{k+1}(t); \\ &\qquad\qquad\qquad\qquad\qquad\qquad k = 0, \ldots, \quad (2) \end{aligned}$$

where $F_k(t)$ is the distribution function of T_k and $F_0(t) = 1$ by definition. From here we can find the average number of renewals in the interval $(0, t]$. That average, which is of course a function of t, is called the 'renewal function'; we shall denote it by $H(t)$:

$$H(t) = E(K_t) = \sum_{k=1}^{\infty} k[F_k(t) - F_{k+1}(t)] = \sum_{k=1}^{\infty} F_k(t). \tag{3}$$

This, again, is a rather difficult expression to evaluate exactly. However, a very simple approximation is available for large values of t, provided that the variance v^2 is finite:

$$H(t) = \frac{t}{m} + \frac{v^2 - m^2}{2m^2} + o(1), \tag{4}$$

where $o(1) \to 0$ when $t \to \infty$. The proof of this result is non-trivial and will be omitted. On the other hand, the fact that the dominant term in $H(t)$ is t/m is intuitively obvious.

From (4) it follows that, in the long run, the average number of renewals during any interval of length y is approximately equal to

$$H(t + y) - H(t) \approx \frac{y}{m}. \tag{5}$$

Suppose now that the distribution $F(x)$ of the renewal periods is continuous, or at least that $F(0) = 0$. Then the probability of more than one renewal occurring in a small interval of length h is $o(h)$ (i.e. negligible compared with h). Hence, the average number of renewals in a small interval $(t, t + h)$ is approximately equal to the probability that there is a renewal in that interval. Thus we have

$$P(\text{a renewal occurs between } t \text{ and } t + h) \approx \frac{h}{m}. \tag{6}$$

In the context of job arrival streams, m is the average interarrival time. Equations (5) and (6) can be summarised in the following rather simple, but nevertheless important, proposition:

Arrival rate lemma. In the long run, the rate of arrivals, i.e. the average number of arrivals per unit time, is constant and is equal to the reciprocal of the average interarrival time, $1/m$.

Another quantity that is of interest in connection with a renewal process is the so-called 'residual life' of the renewal period (also referred to as the 'forward recurrence time'). This is defined as follows: suppose that the renewal process is observed at random over a long run. In other words, an observation point is chosen so that it is equally likely to fall anywhere within a very long interval of time. Let that observation point be T, and the two renewal points between which it falls be T_i and T_{i+1}. The remainder of the observed renewal period, $T_{i+1} - T$, is its residual life (figure 2.1).

Figure 2.1

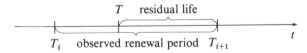

In reliability applications, the residual life is the time between the random observation point and the next machine replacement instant. For an arrival stream, it is the time between the observation point and the next arrival instant.

Denote the residual life by R and its probability density function by $r(x)$. First we remark that if the observed renewal period is of length y, then the observation point is uniformly distributed on $(0, y)$. Therefore, the density function of R, conditioned upon y, is $1/y$ (see section 1.2). Hence, if $s(y)$ is the density function of the observed renewal period (i.e. if the latter has length y with probability $s(y)\mathrm{d}y$), $r(x)$ can be expressed as

$$r(x) = \int_x^\infty s(y) \frac{1}{y} \mathrm{d}y. \tag{7}$$

(This is an application of the complete probability formula (1.9), with the sum replaced by an integral.)

Note that the introduction of $s(y)$ is necessary. We cannot use in (7) the density function of the renewal periods, $f(y)$, because the probability that the observed renewal period is of length y is not, in general, the same as the probability that an arbitrary renewal period is of length y. The observation point is more likely to drop into a large renewal period than into a small one. In fact, the probability that the observed renewal period is of length y is proportional to both y and the relative frequency with which renewal periods of length y occur:

$$s(y) = Cyf(y). \tag{8}$$

The coefficient of proportionality, C, is determined from the fact that $s(y)$, being a probability density function, must integrate to 1:

$$C = \left[\int_0^\infty yf(y)\mathrm{d}y \right]^{-1} = \frac{1}{m}. \tag{9}$$

Substituting (9) and (8) into (7) we obtain

$$r(x) = \frac{1}{m} \int_x^\infty f(y)\mathrm{d}y = \frac{1}{m}[1 - F(x)]. \tag{10}$$

The average residual life is now easily calculated:

$$E(R) = \frac{1}{m} \int_0^\infty x[1 - F(x)]\mathrm{d}x = \frac{1}{2m} \int_0^\infty x^2 f(x)\mathrm{d}x = \frac{M_2}{2m}, \tag{11}$$

where M_2 is the second moment of the renewal period (the second equality in (11) involved an integration by parts).

This last result is somewhat counterintuitive. One's first inclination is to say that, since the observation point is equally likely to fall anywhere and since the average length of a renewal period is m, the average residual life ought to be $m/2$. Instead, we learn from (11) that the average residual life is never less than $m/2$ (the inequality $M_2 \geqslant m^2$ always holds), and may be much much longer. This may lead to apparently paradoxical situations, such as having to wait at a bus stop for 30 minutes, on the average, even though the average interval between bus arrivals is 5 minutes. What is happening here is that, while the overall average is composed of many short intervals and a few long ones, the random observer tends to observe the long intervals and miss the short ones.

Exercises

1. Jobs arrive into a system according to a renewal process, with interarrival periods distributed uniformly on the interval $(0, 2)$. Using the normal approximation, estimate the probability that the time of the 20th arrival will be greater than 25.

2. Derive (11) by first showing that the average length of the observed renewal period is M_2/m, and then arguing that the average residual life must be half of that quantity.

3. The lifetime of a machine is distributed according to the probability density function

$$f(x) = \frac{3}{(1 + x)^4}; \quad x \geq 0.$$

Assuming that the machine is replaced immediately after a breakdown, and that the resulting renewal process is observed at random, find (i) the average lifetime of a machine, (ii) the average lifetime of the observed machine and (iii) the average residual life of the observed machine.

4. Generalise expression (11) by showing that the nth moment of the residual life is given by

$$E(R^n) = \frac{M_{n+1}}{(n + 1)m},$$

where M_n is the nth moment of the renewal period.

5. In the context of a randomly observed renewal process, one may be interested in the time between the observation point and the previous renewal instant ($T - T_i$, in figure 2.1). Show that that random variable, called the 'elapsed life', or the 'backward recurrence time', has the same distribution as the residual life, R.

2.2   The exponential distribution and the Poisson process

Consider now a particular kind of renewal process, namely one where the distribution function of the renewal periods has the form

$$F(x) = 1 - e^{-\lambda x}; \quad x \geq 0. \tag{12}$$

This is called the 'exponential distribution' with parameter λ ($\lambda > 0$); the corresponding probability density function is

$$f(x) = \lambda e^{-\lambda x}; \quad x \geq 0. \tag{13}$$

The parameter of the exponential distribution has a simple physical interpretation. It is the reciprocal of the mean of the corresponding random variable, S:

$$E(S) = \int_0^\infty x\lambda e^{-\lambda x}dx = \frac{1}{\lambda}. \tag{14}$$

A renewal process with exponentially distributed renewal periods is called a 'Poisson process'. As we shall see, these processes are quite remarkable in many ways, and are very widely used in modelling. What makes them special is the following property of the exponential distribution:

The memoryless property. If the random variable S is distributed exponentially, then

$$P(S \leqslant t + s \mid S > t) = P(S \leqslant s); \quad t, s \geqslant 0. \tag{15}$$

In other words, knowing that an exponentially distributed activity has already been in progress for time t does not affect the distribution of its remaining duration; it is as if the activity is starting now. For instance, when interarrival periods are distributed exponentially, the time until the next arrival is independent of the time since the last arrival, and hence of everything that happened in the past. When a machine lifetime is distributed exponentially, the time to a breakdown is independent of how long the machine has been working.

To prove (15), we need only remember the conditional probability formula (1.6):

$$P(S \leqslant t + s \mid S > t) = \frac{P(t < S \leqslant t + s)}{P(S > t)} = \frac{F(t + s) - F(t)}{1 - F(t)}$$

$$= \frac{e^{-\lambda t} - e^{-\lambda(t+s)}}{e^{-\lambda t}} = 1 - e^{-\lambda s} = P(S \leqslant s).$$

Once before we have encountered a memoryless property; that was in the case of the geometric distribution, considered in section 1.4. The analogy with that case extends further: just as the geometric is the only discrete distribution that satisfies (1.49), so the exponential is the only continuous distribution that satisfies (15). The proof of this uniqueness is outlined in exercise 1.

An immediate consequence of either (15) or (10) is that the residual life, R, of an exponentially distributed activity, S, has the same distribution as S. In particular, if S is in progress at time t, then the probability that it will terminate in the interval $(t, t + h)$, where h is small, is independent of t and is approximately equal to λh:

$$P(R \leqslant h) = P(S \leqslant h) = 1 - e^{-\lambda h}$$

$$= 1 - \left[1 - \lambda h + \frac{(\lambda h)^2}{2} - \cdots \right] = \lambda h + o(h). \tag{16}$$

Applying (16) to a Poisson arrival process, where S is an interarrival period, we find that the probability of an arrival in a short interval of length h is approximately λh. This agrees with the arrival rate lemma of the previous section. However, whereas for a general renewal process that result is valid only in the long run (for large values of t), in the Poisson case it holds at all times. In fact, it is possible to take this as a starting point, and define the Poisson process as an arrival process for which

(i) The probability of an arrival in $(t, t + h)$ is equal to $\lambda h + o(h)$, regardless of the process history before time t;

(ii) The probability of more than one arrival in $(t, t + h)$ is $o(h)$, regardless of past history.

The fact that the interarrival periods are exponentially distributed with parameter λ, would then follow from this definition.

From now on, we shall refer to λ as the 'rate' of the Poisson process.

Using the above properties, it is quite easy to find the distribution of the random variable K_t – the number of arrivals in the interval $(0, t)$. That same distribution will also characterise the number of arrivals during any interval of length t (because of the independence of past history).

Divide the interval $(0, t)$ into n sub-intervals of length h each, where $h = t/n$ (figure 2.2).

When h is very small, the probability that there is more than one arrival in one sub-interval is negligible. Then the events 'there are k arrivals in $(0, t)$' and 'k sub-intervals contain an arrival', can be treated as equivalent. Moreover, the occurrence of an arrival in any sub-interval is independent of what happens in the other sub-intervals: these occurrences can be considered as successes in a sequence of Bernoulli trials. Therefore, K_t has approximately the binomial distribution, (1.50), with probability of success λh:

Figure 2.2

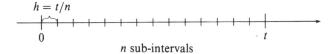

n sub-intervals

$$P(K_t = k) = p_k \approx \binom{n}{k}(\lambda h)^k(1 - \lambda h)^{n-k}.$$

Replacing h with t/n and rearranging terms, this can be re-written as

$$p_k \approx \frac{(\lambda t)^k}{k!}\left(1 - \frac{\lambda t}{n}\right)^n \frac{n(n-1)\cdots(n-k+1)}{n^k}\left(1 - \frac{\lambda t}{n}\right)^{-k}. \tag{17}$$

Now let $n \to \infty$, keeping k fixed. In the limit, the approximate equality (17) will become exact. The first term in the right-hand side of (17) does not depend on n, the second approaches $e^{-\lambda t}$, while the third and fourth approach 1. Thus we find that the distribution of K_t is given by

$$p_k = \frac{(\lambda t)^k}{k!}e^{-\lambda t}; \quad k = 0, 1, \dots. \tag{18}$$

This is called the 'Poisson distribution' with parameter λt. The average number of arrivals in an interval of length t is obtained as

$$E(K_t) = \sum_{k=1}^{\infty} kp_k = \lambda t e^{-\lambda t} \sum_{k=1}^{\infty} \frac{(\lambda t)^{k-1}}{(k-1)!} = \lambda t. \tag{19}$$

Hence the arrival rate λ can also be interpreted as the average number of arrivals per unit time. Again, it is notable that (19) is an exact result, while the analogous expression for general renewal processes, (5), is valid only in the long run.

Consider a Poisson arrival process (rate λ) over two non-intersecting intervals, I_1 and I_2, with lengths t_1 and t_2, respectively. Let I be the union of I_1 and I_2 and K be the number of arrivals that fall in I. Clearly, $K = K_1 + K_2$, where K_j is the number of arrivals during I_j ($j = 1, 2$). The memoryless property implies that K_1 and K_2 are independent random variables whose distributions are not affected by the positions of I_1 and I_2, only by their lengths. Therefore, as far as the distribution of K is concerned, I_1 and I_2 may be assumed adjacent to each other. Then K can be thought of as the number of arrivals during a single interval with length $t_1 + t_2$. Hence, K has the Poisson distribution with parameter $\lambda(t_1 + t_2)$.

The above argument obviously applies to any number of disjoint intervals, thus establishing the following proposition:

The union property. If I is the union of a finite number of disjoint intervals whose lengths sum up to t, and K is the number of arrivals in I from a Poisson process with rate λ, then K has the Poisson distribution with parameter λt.

This property can also be taken as the definition of a Poisson process. It implies both the exponential distribution of the interarrival periods and their independence.

Suppose now that we have two sources of arrivals, say two groups of users submitting jobs to a central computer. The two arrival processes, A_1 and A_2, are independent of each other and are both Poisson with rates λ_1 and λ_2 respectively. What can be said about the merger, or superposition, B, of the two processes (i.e. the process consisting of all arrival instants, regardless of their source)?

Let I be an interval (or a union of a finite number of disjoint intervals) of length t and let K_1, K_2 and K be the numbers of arrivals in I, associated with A_1, A_2 and B, respectively. We know that K_1 and K_2 are independent and their distributions are Poisson, with parameters $\lambda_1 t$ and $\lambda_2 t$ respectively. The distribution of $K = K_1 + K_2$ could be obtained as the convolution of those two distributions. However, the same object can be achieved more simply by observing that, in the expressions (18), λ and t can be interchanged without altering anything. Hence, if K_j ($j = 1,2$) is replaced by the number of arrivals from a Poisson stream with rate t, over an interval of length λ_j, its distribution would be the same. Then K can be regarded as the number of arrivals from a Poisson stream with rate t, over the union of two intervals with total length $\lambda_1 + \lambda_2$. Therefore, K has the Poisson distribution with parameter $(\lambda_1 + \lambda_2)t$. That, in turn, implies that the superposition process B is Poisson, with rate $\lambda_1 + \lambda_2$.

The same argument, applied to the merger of more than two processes, yields:

The superposition property. If $A_1, A_2, \ldots, A_n$ are independent Poisson processes with rates $\lambda_1, \lambda_2, \ldots, \lambda_n$ respectively, then their superposition is also a Poisson process, with rate $\lambda_1 + \lambda_2 + \cdots + \lambda_n$.

The last two properties, together with the central limit theorem, imply that the Poisson distribution approaches the normal one when the parameter λt increases. More precisely, if the random variable K has the Poisson distribution with mean λt (and variance λt; see exercise 2), then the distribution of $(K - \lambda t)/\sqrt{\lambda t}$ approaches the standard normal distribution, $N(0, 1)$, when $\lambda t \to \infty$.

The approximation can simplify considerably the computational task involved in evaluating Poisson probabilities. For example, consider a computing system where jobs arrive according to a Poisson process, at an average rate of 2 jobs per second. Suppose that we are interested in the number of jobs, K, that arrive during a two-minute period, and wish to find the probability that that number does not exceed 260. Here we have $\lambda = 2$, $t = 120$ and $\lambda t = 240$. Using the normal approximation, we get

$$P(K \leqslant 260) = P\left(\frac{K - 240}{\sqrt{240}} \leqslant \frac{20}{\sqrt{240}}\right) \approx F(1.29) \approx 0.90,$$

where $F(x)$ is the standard normal distribution function defined in (1.80). That function is extensively tabulated.

By contrast, an attempt to obtain the above probability by substituting $\lambda t = 240$ in (18), and summing for $k = 0, 1, \ldots, 260$, would be almost certainly doomed to failure. There would be great numerical difficulties arising from floating point overflows and underflows, round-off errors, etc.

For practical purposes, the normal approximation can be applied successfully if λt is greater than about 20.

Consider now the operation of splitting, or decomposing, a Poisson arrival process A, with rate λ, into two arrival processes, B_1 and B_2. For instance, a computer centre may have two processors capable of executing the jobs submitted by its users; some of the incoming jobs could be directed to the first processor and others to the second.

The decomposition is performed by a sequence of Bernoulli trials: every arrival of the process A is assigned to the process B_1 with probability q_1, and to B_2 with probability q_2 $(q_1 + q_2 = 1)$, regardless of all previous assignments. Let K, K_1 and K_2 be the numbers of arrivals associated with A, B_1 and B_2 respectively, during an interval (or a set of disjoint intervals) of length t. We know that K, which is equal to $K_1 + K_2$, has the Poisson distribution with parameter λt; the problem is to find the joint distribution of K_1 and K_2. Bearing in mind that, if the value of K is given, then K_1 and K_2 are the numbers of successes and failures in the corresponding set of Bernoulli trials, we can write

$$
\begin{aligned}
P(K_1 &= k_1, K_2 = k_2) \\
&= P(K_1 = k_1, K_2 = k_2 \mid K = k_1 + k_2) P(K = k_1 + k_2) \\
&= \frac{(k_1 + k_2)!}{k_1! \, k_2!} q_1^{k_1} q_2^{k_2} \frac{(\lambda t)^{k_1 + k_2}}{(k_1 + k_2)!} e^{-\lambda t} \\
&= \frac{(q_1 \lambda t)^{k_1}}{k_1!} e^{-q_1 \lambda t} \frac{(q_2 \lambda t)^{k_2}}{k_2!} e^{-q_2 \lambda t}.
\end{aligned}
\tag{20}
$$

This equation implies that the distributions of K_1 and K_2 are Poisson, with parameters $q_1 \lambda t$ and $q_2 \lambda t$ respectively. Moreover, those random variables are independent of each other. An obvious generalisation of the above argument yields:

The decomposition property. If a Poisson process, A, with rate λ, is decomposed into processes $B_1, B_2, \ldots, B_n$, by assigning each arrival in A to B_i with probability q_i $(i = 1, 2, \ldots, n; \quad q_1 + q_2 + \cdots + q_n = 1)$, independently of all previous assignments, then $B_1, B_2, \ldots, B_n$ are Poisson processes with rates $q_1 \lambda, q_2 \lambda, \ldots, q_n \lambda$ respectively, and are independent of each other.

The superposition and decomposition of Poisson processes are illustrated in figure 2.3.

Figure 2.3

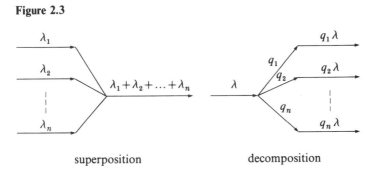

superposition   decomposition

When studying the behaviour of a system where jobs arrive at random, we shall sometimes wish to look at its state 'through the eyes' of the new arrivals. To be precise, suppose that the system state at time t is described by the random variable S_t (i.e. the system is modelled by the stochastic process $\{S_t; t \geqslant 0\}$). If t_0 is a moment when an arrival occurs, then the state just before that moment, $S_{t_0^-}$, is the state 'seen' by the arriving job. Now, obviously the fact that there is an arrival as t_0 influences S_{t_0} and the subsequent states, $S_t, t \geqslant t_0$. However, does that fact influence the state seen by the new arrival, $S_{t_0^-}$?

In general, it does. However, if jobs arrive into the system according to a Poisson process, then the distribution of $S_{t_0^-}$ is independent of whether there is an arrival at t_0 or not. In other words, the system state seen by an arrival from a Poisson process has the same distribution as the state seen by a 'random observer': someone who just happens to look at the system, having otherwise nothing to do with it. This is referred to as the 'random observer property' of the Poisson process. We shall offer the following argument in justification of that property:

The system state seen at time t_0, $S_{t_0^-}$, is influenced only by the arrivals that occur before t_0. When the arrival process is Poisson, the interarrival periods are exponentially distributed and have the memoryless property, looking both forward and backward in time. Hence, the interval between t_0 and the nearest preceding arrival has the same distribution regardless of whether t_0 is an arrival instant or not. This implies that a job arriving at t_0 has no more information about the previous arrivals than a random observer at t_0. Therefore, both see the same distribution of the system state.

It will be instructive, perhaps, to give an example of a system where jobs do not arrive according to a Poisson process and do not have the random observer property. Suppose that data packets are submitted for transmission by a communication channel at intervals which range between 20 and 30 seconds. Each transmission lasts between 10 and 15 seconds. Then an arriving packet is guaranteed to see the channel idle, whereas a random observer may well see it busy.

Given all the nice properties of the Poisson process, it is hardly surprising that the analysis of a system model is greatly facilitated when one can assume that the stream of service requests arriving into the system is Poisson. Fortunately, such an assumption can often be justified on grounds other than those of analytical convenience. The Poisson distribution, like the normal one, turns up as the limiting distribution of large sums.

The following result holds under some mild restrictions:

The superposition of n arbitrary, independent renewal processes, each with average renewal period n/λ, approaches a Poisson process with rate λ when n increases.

Thus, if the requests for service arriving at a computer installation, or a communication channel, are submitted by many independent users, and if each user's requests are reasonably widely spaced, then the overall arrival process will be approximately Poisson.

Exercises

1. Show that if a continuous non-negative random variable S satisfies (15), then S is exponentially distributed. Hint: relation (15) implies that the distribution function of S, $F(x)$, satisfies a differential equation of the form $F'(x) = \lambda[1 - F(x)]$, for some $\lambda > 0$. Hence, $F(x) = 1 - e^{-\lambda x}$.

2. The variance of the number of arrivals during an interval of length t, for a Poisson process with rate λ, is given by $\text{Var}(K_t) = \lambda t$. Demonstrate this either by manipulating the infinite series resulting from (18) or, more simply, by the approach illustrated in figure 2.2, using the known variance of the binomial distribution and taking the limit $n \to \infty$.

3. A certain computer system is subject to crashes which occur according to a Poisson process, at the rate of 10 per year (assume, for simplicity, that repairs are instantaneous). Given that there was exactly one crash during the month of January, what is the probability that it occurred on the last day of that month?

4. Compilation jobs arrive at a computer centre according to a Poisson

process, at the rate of 3 jobs per minute. The compiler requested in each case is Pascal with probability 0.6, C with probability 0.3 and Simula with probability 0.1. What is the probability that the numbers of Pascal, C and Simula compilations requested during an interval of one hour will exceed 100, 50 and 20, respectively? Hint: use the decomposition property and the normal approximation.

5. Let T_n be the instant of the nth arrival in a Poisson process with rate λ. Show that the distribution function of T_n, $F_n(x)$, is given by

$$F_n(x) = 1 - \sum_{k=0}^{n-1} \frac{(\lambda x)^k}{k!} \, e^{-\lambda x}.$$

This is called the 'Erlang distribution function'. The probability density function corresponding to it is

$$f_n(x) = F'_n(x) = \frac{\lambda(\lambda x)^{n-1} e^{-\lambda x}}{(n-1)!}.$$

(Hint: use the fact that $P(T_n \leqslant x) = P(K_x \geqslant n)$.)

2.3 Markov chains

In this, and in the following section, we shall examine a class of stochastic processes which are characterised by what is known as the 'Markov property'. In general terms, this can be stated as follows:

If the state of the process at a given moment in time is known, then its subsequent behaviour is independent of its past history.

These 'Markov processes' (named after the Russian mathematician who first studied them) provide us with one of the most important tools of modelling. The Markov property strikes that happy balance between complexity and simplicity which make it, on one hand, sufficiently general to be useful and, on the other, sufficiently special to be susceptible to analysis.

The Markov processes which are of interest to us are the discrete state ones. That is, the process states can be numbered by the set, or by a subset, of the non-negative integers $\{0, 1, \ldots\}$. We shall denote the state space by S. If, in addition, the time parameter is also discrete, i.e. if the process is observed at time instants $0, 1, \ldots$, then we say that we are dealing with a 'Markov chain'. By implication, the term 'Markov process' is reserved for the continuous time case.

Let $X = \{X_n; n = 0, 1, \ldots\}$ be a Markov chain. The Markov property says, in this case, that given the state of X at time n, its state at time $n + 1$ is independent of the states at times $0, 1, \ldots, n - 1$:

$$P(X_{n+1} = j \mid X_0, X_1, \ldots, X_n) = P(X_{n+1} = j \mid X_n);$$
$$j, n = 0, 1, \ldots \quad (21)$$

Hence, the evolution of the Markov chain is completely described by the 'one-step transition probabilities', $q_{ij}(n)$, that the chain will move to state j at time $n + 1$, given that it is in state i at time n:

$$q_{ij}(n) = P(X_{n+1} = j \mid X_n = i); \quad i, j, n = 0, 1, \ldots \quad (22)$$

From now on, we shall assume that the one-step transition probabilities do not depend on the time instant:

$$q_{ij}(n) = q_{ij}; \quad i, j, n = 0, 1, \ldots \quad (23)$$

This restriction simplifies the treatment, without reducing the generality of the theory. Markov chains which satisfy (23) are said to be 'time-homogeneous'. That qualification will not be mentioned in future, but will be implied.

Thus, a Markov chain is characterised by its 'transition probability matrix', Q, given by

$$Q = \begin{bmatrix} q_{00} & q_{01} & q_{02} & \cdots \\ q_{10} & q_{11} & q_{12} & \cdots \\ \vdots & \vdots & \vdots & \cdots \\ q_{i0} & q_{i1} & q_{i2} & \cdots \\ \vdots & \vdots & \vdots & \cdots \end{bmatrix}, \quad (24)$$

where the indices range over the state space, S. The elements in each row of the transition probability matrix sum to 1 (after state i, the chain must find itself in some state):

$$\sum_{j \in S} q_{ij} = 1; \quad i = 0, 1, \ldots \quad (25)$$

Conversely, for any square matrix, Q (finite or infinite), whose elements are non-negative and satisfy (25), one can construct a Markov chain which has Q as its transition probability matrix.

A very convenient and visually appealing representation of a Markov chain is obtained by drawing a directed graph, with vertices corresponding to the states of the chain and arcs showing the non-zero one-step transition probabilities. That graph is called the 'state diagram' of the Markov chain.

Examples

1. A telephone line can be in one of two possible states: idle or busy (0 or 1, respectively). If it is idle at time n, it will be idle or busy at time $n + 1$ with probabilities 0.9 and 0.1 respectively. If it is busy at time n, it will be idle or busy at time $n + 1$ with probabilities 0.3 and 0.7 respectively. This behaviour is modelled by a two-state Markov chain with transition probability matrix

$$Q = \begin{bmatrix} 0.9 & 0.1 \\ 0.3 & 0.7 \end{bmatrix}.$$

The state diagram of that Markov chain is shown in figure 2.4.

Figure 2.4

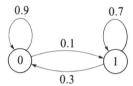

2. An input/output buffer has room for M records. In any one unit of time, a new record may be inserted into the buffer (provided that the latter is not full), with probability α. Also, in any unit interval, the buffer may be emptied completely, with probability β. If both an insertion and a clearing operation occur in the same unit interval, the former is carried out before the latter.

Let X_n be the number of messages in the buffer at time n. Assuming that the occurrences of insertions and clearings are independent of each other and of their past histories, $\{X_n; n = 0, 1, \ldots\}$ is a Markov chain with a state space $\{0, 1, \ldots, M\}$; its state diagram is illustrated in figure 2.5.

Figure 2.5

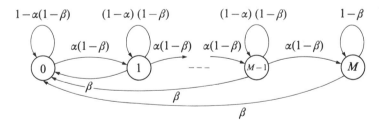

3. A gambler bets £1 at a time on the toss of a die. He loses his stake if the number that comes up is less than 5, and wins £2 if it is 5 or 6. He stops playing when his capital reaches either £0 or £100.

If X_0 is the gambler's initial capital and X_n is the amount he has immediately following the nth move, $\{X_n; n = 0, 1, \ldots\}$ is a Markov chain. The transition probabilities have the following values: $q_{ii} = 1$ if $i = 0$ or $i \geqslant 100$; $q_{i,i-1} = 2/3$, $q_{i,i+2} = 1/3$ for $i = 1, 2, \ldots, 99$; $q_{ij} = 0$ otherwise.

4. A population of single-cell organisms evolves according to the following simple rules: the lifespan of every organism is exactly one unit of time. This terminates either with the death of the organism, with probability γ, or with its replacement by two new ones, with probability $1 - \gamma$, independently of all the others.

If X_n is the size of the population at time n (X_0 being the initial size), then $\{X_n; n = 0, 1, \ldots\}$ is a Markov chain whose state space is the set of all non-negative integers. It is easy to see, however, that although X_0 can have an arbitrary value in that set, the values of $X_1, X_2, \ldots$ are always even. This is because the only possible one-step transitions from state i ($i = 1, \ldots$) are to states $0, 2, 4, \ldots, 2i$. The corresponding transition probabilities are

$$q_{i,2j} = \binom{i}{j}\gamma^{i-j}(1-\gamma)^j; \quad j = 0, 1, \ldots, i$$

(j of the i organisms divide in two and the other $i - j$ die).

Let us consider now the transitions that a · Markov chain $\{X_n; n = 0, 1, \ldots\}$ can make in two steps, rather than one. Because of the time-homogeneity, the probabilities of those transitions are also independent of the step index. We shall use the notation

$$q_{ij}^{(2)} = P(X_{n+2} = j \,|\, X_n = i); \quad i, j, n = 0, 1, \ldots. \tag{26}$$

Since, in order to move from state i to state j in two steps, the chain has to pass through some intermediate state, k, after the first step, we can write

$$q_{ij}^{(2)} = \sum_{k \in S} P(X_{n+1} = k, X_{n+2} = j \,|\, X_n = i)$$

$$= \sum_{k \in S} P(X_{n+2} = j \,|\, X_n = i, X_{n+1} = k) P(X_{n+1} = k \,|\, X_n = i)$$

$$= \sum_{k \in S} P(X_{n+2} = j \,|\, X_{n+1} = k) P(X_{n+1} = k \,|\, X_n = i)$$

$$= \sum_{k \in S} q_{ik} q_{kj} \tag{27}$$

(the third equality in (27) uses the Markov property).

Note that the right-hand side of (27) is equal to the (i, j)th element of the square of the transitions probability matrix, Q^2. A simple inductive extension of this argument leads to the following general proposition:

The m-step transition probabilities of a Markov chain, defined as

$$q_{ij}^{(m)} = P(X_{n+m} = j \mid X_n = i); \quad m = 1, 2, \dots, \tag{28}$$

are the elements of the mth power of the transition probability matrix, Q^m.

This result may, in principle, be used to compute various quantities of interest associated with the performance of a Markov chain over a finite period of time. For instance, in example 2, we may wish to find the average number of messages in the buffer at time 50, given that at time 0 the buffer was empty. The appropriate expression is

$$E(X_{50} \mid X_0 = 0) = \sum_{j=1}^{M} j q_{0j}^{(50)}$$

(in fact, the upper limit of the summation can be replaced by $\min(50, M)$, since there cannot be more than 50 messages in the buffer at time 50).

It should be pointed out, however, that in most cases of practical interest, evaluating powers of the transition probability matrix is prohibitively expensive. As a consequence, problems concerned with the short-term and medium-term performance of a system are often numerically intractable.

Analysing the long-term behaviour of a Markov chain tends to be simpler. This is because, under certain conditions, the more transitions the chain goes through, the less it matters in what state it was when it started. In the limit, when the moment of observation is infinitely far removed from the starting point, the probability, p_j, of finding the chain in state j ($j = 0, 1, \dots$), is independent of the initial state:

$$p_j = \lim_{n \to \infty} P(X_n = j \mid X_0 = i) = \lim_{n \to \infty} q_{ij}^{(n)}; \quad j = 0, 1, \dots. \tag{29}$$

When the limiting probabilities p_j exist, and add up to 1, they are referred to as the 'long-run distribution', or the 'steady-state distribution', or the 'equilibrium distribution' of the Markov chain. The problems of deciding whether a steady-state distribution exists, and of determining it if it does, are central to both the theory and the applications of Markov chains. We shall state some of the important results in this area, but in order to do that, a few definitions will be necessary.

Denote by v_{ij} the probability that, having been in state i, the chain will eventually reach state j (this does not depend on when the visit to i occurred):

$$v_{ij} = P(X_n = j \text{ for some } n \mid X_0 = i); \quad i, j = 0, 1, \dots. \tag{30}$$

If v_{ij} is not equal to 0, then state j is said to be 'reachable' from state i. An equivalent, and easier to use, definition of the same concept, can be given by means of the state diagram: if, starting in state i and following the transition arrows from state to state, one can arrive at state j, then j is reachable from i. This relation is not necessarily reflexive: j may be reachable from i without i being reachable from j.

In examples 1 and 2, every state is reachable from every other state. In example 3, if $1 \leqslant i \leqslant 99$ and $0 \leqslant j \leqslant 100$, then j is reachable from i; from state 0, only state 0 is reachable and similarly for state 100. Such states are called 'absorbing'. In example 4, state 0 is absorbing and every even-numbered state is reachable from every state.

A set, C, of states is said to be 'closed' if no state outside C can be reached from a state in C. The set of all states is obviously closed. If that is the smallest closed set of states, i.e. if no proper subset of the set of all states is closed, then the Markov chain is said to be 'irreducible'. It is readily seen that a Markov chain is irreducible if, and only if, every state can be reached from every other state. The chains in examples 1 and 2 are irreducible, but those in examples 3 and 4 are not.

If $v_{ii} = 1$ for some state i, then having once visited state i the chain is certain to return to it eventually. Such states are called 'recurrent'. Clearly, if i is a recurrent state, then it is visited by the Markov chain either not at all, or infinitely many times (since no visit to i can be the last one).

States which are not recurrent are called 'transient'. It is not difficult to show that the total number of visits to a transient state is finite, with probability 1 (see exercise 2). Every transient state is eventually visited by the Markov chain for the last time, and then no more. For instance, if C is a closed set of states and i is not in C, and if a state in C is reachable from i, then i is transient. With a non-zero probability, after visiting state i the chain will enter C and will not afterwards return to i. Such is the situation in example 3: the states $1, 2, \ldots, 99$ are transient. In the long run, the Markov chain will be trapped in one of the absorbing states, 0 or $j \geqslant 100$.

Clearly, the long-run probability of any transient state is zero. If i is a recurrent state and j is a state reachable from i, then j is recurrent too. Indeed, suppose that j is transient and consider the path followed by the chain after a visit to j. A return to j is not certain. Hence, a visit to i is also not certain (if it was, then the chain would keep on visiting i and would eventually find its way back to j). That means, however, that it is possible for the chain to move from i to j and then not return to i, which contradicts the fact that i is recurrent. Therefore, j is recurrent.

The above argument implies that, if a Markov chain is irreducible, then either all its states are transient or all its states are recurrent. We refer to these two cases by saying that the chain itself is transient, or recurrent, respectively.

Consider the operation path of an irreducible and recurrent Markov chain $X = \{X_n; n = 0, 1, \ldots\}$. Every state is visited by X infinitely many times, with probability 1, regardless of the initial state. Let $n_1^j, n_2^j, \ldots$ be the moments of consecutive visits to state j. Denote by m_j the expected interval between those visits:

$$m_j = E(n_{k+1}^j - n_k^j); \quad k = 1, 2, \ldots . \tag{31}$$

Since an average of one out of m_j observations finds X in state j, our Markov chain spends a fraction $1/m_j$ of its time in state j. Intuitively, that fraction should be equal to the long-run probability, p_j, of observing X in state j. This is indeed true, provided that the pattern of visits to the various states is not too regular.

A state j is said to be 'periodic', with period m ($m > 1$), if the consecutive returns to j can only occur at multiples of m steps:

$$P(X_{n+km} = j \text{ for some } k \geqslant 1 \,|\, X_n = j) = 1. \tag{32}$$

If there is no integer $m > 1$ which satisfies (32), j is said to be 'aperiodic'. When the Markov chain is irreducible, it can be shown that either all its states are periodic, with the same period, or all of them are aperiodic. The chain itself is then called periodic, or aperiodic, respectively.

The first limiting result can now be stated:

If $X = \{X_n; n = 0, 1, \ldots\}$ is an irreducible, aperiodic and recurrent Markov chain, then the limiting probabilities p_j, defined in (29), exist and are given by

$$p_j = \frac{1}{m_j}; \quad j = 0, 1, \ldots . \tag{33}$$

We shall omit the proof. Note that, although X keeps returning to state j with probability 1, the average intervals between those returns may be infinitely large. Such states are called 'recurrent null'; their limiting probabilities are equal to 0. The states whose average return times are finite, and whose long-run probabilities are therefore non-zero, are called 'recurrent non-null'. Again it can be shown that either all states of an irreducible Markov chain are recurrent non-null, or none of them is.

Relation (33) is quite instructive, but it does not tell us how to decide whether the states of X are recurrent non-null. Nor does it provide an algorithm for determining the probabilities p_j in terms of the parameters of X. What is really needed is a connection between the one-step transition probabilities of a Markov chain and its steady-state distribution. This is supplied by the following result:

Steady-state theorem for Markov chains. An irreducible and aperiodic

Markov chain, X, with state space S and one-step transition probability matrix $Q = (q_{ij})$, $i, j \in S$, is recurrent non-null if, and only if, the system of equations

$$p_j = \sum_{i \in S} p_i q_{ij}; \quad j = 0, 1, \ldots, \tag{34}$$

$$\sum_{j \in S} p_j = 1, \tag{35}$$

has a solution. That solution is then unique and is the steady-state distribution of X.

Equations (34) are referred to as the 'balance equations' of the Markov chain X, while (35) is the normalising equation.

The balance equations have a simple intuitive interpretation which makes their necessity for the existence of a steady-state distribution almost obvious. Indeed, suppose that the limits (29) exist, and consider the chain at some moment in the steady state (i.e. after it has been running for a long time). At that moment, the chain is in state i with probability p_i and, if that is the case, the next state entered will be j with probability q_{ij}. Hence, the right-hand side of (34) is equal to the unconditional probability that at the next observation instant the chain will be in state j. But the next observation instant is also in the steady state, which means that that probability must be equal to p_j.

To show the sufficiency of (34) and (35), let the initial state of the Markov chain be chosen at random, with the distribution p_j ($j = 0, 1, \ldots$). Then, since the balance equations are satisfied, the same distribution will be preserved at all subsequent observation instants, no matter how distant. That, however, implies that the Markov chain can be neither transient, nor recurrent null, for in both those cases all the long-run probabilities would be 0, contradicting (35).

The first limiting result, (33), tells us that a Markov chain cannot have more than one steady-state distribution. From this, and from the above argument, it follows that the solution to (34) and (35), if it exists, is unique.

Thus, the steady-state analysis of a system modelled by an irreducible and aperiodic Markov chain reduces to the solution of the corresponding balance and normalising equations. That may be an easy or a difficult task, depending on the size of the state space and on the structure of the one-step transition probability matrix.

When the state space is finite, the chain is always recurrent non-null (see exercise 5) and therefore (34) and (35) always have a solution. Numerically, that solution is usually obtained by discarding one of the balance equations and replacing it with the normalising equation.

Let us apply this approach to the Markov chains in examples 1 and 2 in this section. Both those chains are irreducible, because all their states can be reached from each other. Moreover, both chains are aperiodic, since they have states to which they can return after a single step.

The balance equations for the Markov chain modelling the state of a telephone line are

$$p_0 = 0.9p_0 + 0.3p_1,$$
$$p_1 = 0.1p_0 + 0.7p_1.$$

Solving the first of these, say, together with the normalising equation $p_0 + p_1 = 1$, yields $p_0 = 0.75$, $p_1 = 0.25$. In other words, the telephone line will be busy for a fraction $1/4$ of the time, in the long run.

Using the state diagram of the input/output buffer we can easily write the appropriate balance equations. The right-hand side of the jth equation $(j = 0, 1, \ldots, M)$ contains the long-run probabilities of the states from which there are arrows leading to state j, multiplied by the corresponding one-step transition probabilities:

$$p_0 = [1 - \alpha(1 - \beta)]p_0 + \beta(p_1 + p_2 + \cdots + p_M),$$
$$p_1 = \alpha(1 - \beta)p_0 + (1 - \alpha)(1 - \beta)p_1,$$
$$\vdots$$
$$p_{M-1} = \alpha(1 - \beta)p_{M-2} + (1 - \alpha)(1 - \beta)p_{M-1},$$
$$p_M = \alpha(1 - \beta)p_{M-1} + (1 - \beta)p_M.$$

From equations $1, 2, \ldots, M$ we obtain, by consecutive elimination

$$p_j = \gamma^j p_0; \quad j = 0, 1, \ldots, M - 1; \quad p_M = \frac{\alpha(1 - \beta)}{\beta} \gamma^{M-1} p_0,$$

where

$$\gamma = \frac{\alpha(1 - \beta)}{1 - (1 - \alpha)(1 - \beta)}.$$

The normalising equation then yields

$$p_0 = \frac{\beta\gamma}{\alpha(1 - \beta)} = \frac{\beta}{1 - (1 - \alpha)(1 - \beta)},$$

thus determining all the steady-state probabilities. From these, one can derive various performance characteristics. For instance, the steady-state average number of messages in the buffer, L, is given by

$$L = \sum_{j=1}^{M} j p_j.$$

Exercises

1. Assuming that the gambler in example 3 starts playing with an initial capital of £2, what amounts could he have after three moves, and with what probabilities?

2. Let j be a transient state for some Markov chain, with probability v_{jj} that, after visiting j, the chain will eventually return to it. Show that, if the chain visits j at all, then the total number of such visits, K_j, is geometrically distributed with parameter v_{jj}:

$$P(K_j = k) = v_{jj}^{k-1}(1 - v_{jj}); \quad k = 1, 2, \ldots .$$

Hence deduce that K_j is finite, with probability 1.

3. Consider a modified version of the input/output (I/O) buffer of example 2, where the clearing operation is replaced by one which removes a single message from the buffer (if the latter is not empty). All other assumptions, and the parameters, remain as before. Draw the state diagram of the new Markov chain. Show that this chain, too, is irreducible and aperiodic, and find its steady-state distribution.

4. (Random walk). A particle meanders among the non-negative integer points on the x-axis. If, at time n, it is in position i ($i = 1, 2, \ldots$), then at time $n + 1$ it can move to point $i - 1$ with probability α, or to point $i + 1$ with probability $1 - \alpha$ ($0 < \alpha < 1$). If it is at point 0, then it either remains there, with probability α, or moves to point 1 with probability $1 - \alpha$.

Draw the state diagram of the Markov chain $X = \{X_n; n = 0, 1, \ldots\}$, where X_n is the position of the particle at time n. Show that X is irreducible and aperiodic, and that the solution of the balance equations is of the form $p_j = \gamma^j p_0$, where $\gamma = (1 - \alpha)/\alpha$. Hence demonstrate that X is recurrent non-null if, and only if, $\gamma < 1$, or $\alpha > 1/2$. In that case, the steady-state distribution of X is geometric:

$$p_j = \gamma^j(1 - \gamma); \quad j = 0, 1, \ldots .$$

5. Consider an irreducible Markov chain $X = \{X_n; n = 0, 1, \ldots\}$, with a finite state space $S = \{0, 1, \ldots, M\}$. Assume that X is transient, or recurrent null, and that the limiting probabilities of all states are therefore equal to 0. Show that this leads to a contradiction with the identity

$$\sum_{i=0}^{M} P(X_n = i) = 1,$$

which holds for every n, regardless of the initial state. Hence deduce that X is recurrent non-null.

2.4　　Markov processes

We now turn our attention to the continuous time analogue of the Markov chain. This is a stochastic process, $X = \{X_t; t \geqslant 0\}$, whose time parameter takes arbitrary non-negative real values. The state space, S, is still assumed to be discrete; it can be identified with the set, or a subset, of the non-negative integers.

In most systems of practical interest, changes of state can occur at arbitrary moments in time, and the intervals between those changes can be of arbitrary length. When that is the case, a model based on a continuous time process is in many respects preferable to a discrete time one.

The process X is called a 'Markov process' if it has the Markov property. That is, the path followed by X after a given moment, t, depends only on the state at that moment, X_t, and not on the past history:

$$P(X_{t+s} = j \mid X_u; u \leqslant t) = P(X_{t+s} = j \mid X_t);$$
$$j = 0, 1, \ldots ; s, t \geqslant 0. \quad (36)$$

The theory of Markov processes, especially as far as their long-term behaviour is concerned, has many parallels with that of Markov chains. In particular, much of the terminology of the previous section carries over almost unchanged.

A Markov process X is said to be 'time-homogeneous' if the right-hand side of (36) does not depend on the moment of observation, t, but is a function only of s, j and the value of X_t:

$$P(X_{t+s} = j \mid X_t = i) = q_{ij}(s); \quad i, j = 0, 1, \ldots ; s \geqslant 0. \quad (37)$$

From now on, all Markov processes will be assumed time-homogeneous.

The functions (37) are called the 'transition probability functions' of the Markov process. They are analogous to the m-step transition probabilities of a Markov chain, defined in (28).

The evolution of a typical Markov process (excluding certain pathological cases) can be described as follows:

The process enters a state, say i $(i = 0, 1, \ldots)$. It remains there for a random period of time, distributed exponentially with some parameter, μ_i. At the end of that period, the process moves to a different state, j $(j = 0, 1, \ldots; j \neq i)$, with some probability, q_{ij}. It then remains in state j for a period of time distributed exponentially with parameter μ_j, moves to state k with probability q_{jk}, etc.

This sort of behaviour is implied by the Markov property. Indeed, the latter tells us that, if at any moment the process is observed in state i, the time that it will remain in that state is independent of the time already spent in it. But we know (see exercise 1 of section 2.2) that the only distribution which has this memoryless property is the exponential one. Similarly, the next state to be entered depends only on the current state, and not on the time spent in it or on any previous states.

The significance of the above process structure is, for us, two-fold. First, the Markov property and the exponential distribution are very closely related. To assume that a system can be modelled by a Markov process is, in essence, equivalent to assuming that the intervals between events which change the system state are exponentially distributed.

The second important observation is that a Markov process is completely characterised by the parameters μ_i and the transition probabilities q_{ij} $(i, j = 0, 1, \ldots; i \neq j)$. In fact, we shall see that it is completely characterised by the products

$$a_{ij} = \mu_i q_{ij}; \quad i, j = 0, 1, \ldots; i \neq j. \tag{38}$$

These quantities are called the 'generators', or the 'instantaneous transition rates' of the Markov process. The latter name can be justified as follows:

Suppose that our process is in state i at time t, and let h be a small time increment. What is the probability that the process will be in state j at time $t + h$? For that to happen, the period of residence in state i must terminate within the interval h and the process must move to state j (figure 2.6).

Figure 2.6

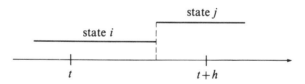

Remembering equation (16), we can write

$$P(X_{t+h} = j \mid X_t = i) = \mu_i h q_{ij} + o(h) = a_{ij} h + o(h); \qquad i \neq j, \tag{39}$$

where $o(h)$ is a term which is negligible, compared with h (that term takes into account multiple transitions, such as from i to k and then from k to j, within the small interval h).

Thus a_{ij} is the instantaneous rate at which the process moves from state i to state j. Alternatively, a_{ij} can be interpreted as the average number of transitions from state i to state j, per unit time spent in state i.

The instantaneous transition rates are the parameters of a Markov process model. They are either given as part of the model formulation, or are obtained from other basic assumptions. The object of any analysis is to derive performance measures in terms of these parameters.

Examples

1. Consider a Poisson arrival process with rate λ, and let K_t be the number of arrivals in the interval $(0, t)$. Then $K = \{K_t; \, t \geqslant 0\}$ is clearly a Markov process, since the instants of future arrivals do not depend on those of past ones. If the process is in state i at time t, it will be in state $i + 1$ at time $t + h$ if one new arrival occurs between t and $t + h$; the probability of that event is $\lambda h + o(h)$ (see section 2.2). The probability of more than one arrival in an interval of length h is $o(h)$. Hence, the instantaneous transition rates are given by

$$a_{ij} = \begin{cases} \lambda & \text{if } j = i + 1, \\ 0 & \text{if } j \neq i, \, i + 1. \end{cases}$$

2. (Continuous time version of the I/O buffer from example 2, section 2.3). Records arrive into a buffer in a Poisson stream with rate λ. There is room for M records; any new arrival which finds the buffer full is lost. From time to time the buffer is cleared completely. The intervals between those moments are independent of the arrivals and are distributed exponentially with parameter μ (i.e. the clearing instants occur in a Poisson stream with rate μ).

Let X_t be the number of records in the buffer at time t. The above assumptions ensure that $X = \{X_t; \, t \geqslant 0\}$ is a Markov process whose state space is the set $\{0, 1, \ldots, M\}$. If X is in state i at time t, the states to which it can move by the time $t + h$ with a non-negligible probability are $i + 1$ (if $i < M$ and a new record arrives between t and $t + h$) or 0 (if $i > 0$ and a clearing instant occurs between t and $t + h$). The probabilities of those transitions are $\lambda h + o(h)$ and $\mu h + o(h)$ respectively. The probability of more than one arrival, or of an arrival and a clearing instant in a small interval of length h, is $o(h)$. Hence, the non-zero instantaneous transition rates of X are

$$a_{i,i+1} = \lambda, \quad i = 0, 1, \ldots, M - 1,$$
$$a_{i0} = \mu, \quad i = 1, 2, \ldots, M.$$

3. A small telephone network connects four subscribers. Each of them attempts to make calls at intervals which are distributed exponentially with parameter λ, independently of the others. The recipient of a call is equally likely to be any of the other three subscribers, regardless of past history. If that recipient is not busy at the time, the call is successful and a conversation begins; otherwise the call is lost. The durations of conversations are distributed exponentially with parameter μ.

As a state descriptor, we can use the number of conversations in progress at time t, X_t. Because of the memoryless assumptions, $X = \{X_t; t > 0\}$ is a Markov process with a state space $\{0, 1, 2\}$. Let us consider the possible transitions out of each of these three states.

In state 0, there are four unoccupied subscribers, each of which can make a call in a small interval of length h with probability $\lambda h + o(h)$. If anyone does, the call will be successful, since the recipient is known to be free. Hence the probability that the process will move to state 1 is $4\lambda h + o(h)$. In state 1, there is one conversation in progress and two free subscribers. An attempt will be made to make a call in $(t, t + h)$ with probability $2\lambda h + o(h)$. However, such an attempt will be successful only if it is addressed to the other free subscriber, which will happen with probability $1/3$. Hence, the process will move to state 2 with probability $(2/3)\lambda h + o(h)$. It can also move to state 0, if the conversation which is in progress terminates. This will happen with probability $\mu h + o(h)$. In state 2, there are two conversations in progress and no free subscribers. Since either conversation can terminate, with probability $\mu h + o(h)$, the probability that the process will move to state 1 is $2\mu h + o(h)$. All other transitions occur with probability $o(h)$.

Thus, the instantaneous transition rates for our process are:

$$a_{0,1} = 4\lambda; \quad a_{1,2} = \tfrac{2}{3}\lambda; \quad a_{1,0} = \mu; \quad a_{2,1} = 2\mu.$$

Just as in the case of Markov chains, it is very convenient and useful to describe a Markov process by means of a state diagram. This is, once again, a directed graph where the nodes represent the process states. The arcs are now labelled with the corresponding instantaneous transition rates. The state diagrams for examples 2 and 3 are shown in figure 2.7, (a) and (b) respectively.

Figure 2.7

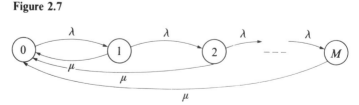

(a) Buffer with room for M records

Figure 2.7 (*continued*)

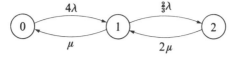

(b) Telephone network with four subscribers

Let us return now to the definitions (38) of the instantaneous transition rates. Bearing in mind that the sum of the transition probabilities q_{ij} over all j $(j \neq i)$ must be 1, we can express the parameter μ_i in terms of a_{ij}:

$$\mu_i = \sum_{\substack{j \in S \\ j \neq i}} a_{ij}; \quad i = 0, 1, \ldots . \tag{40}$$

We can think of μ_i as the instantaneous transition rate out of state i.

So far, the quantities a_{ij} have not constituted a matrix, because they have been defined only for $i \neq j$. We now remedy this situation by defining $a_{ii} = -\mu_i$ $(i = 0, 1, \ldots)$, and forming the 'generator matrix', A:

$$A = \begin{bmatrix} a_{00} & a_{01} & a_{02} & \cdots \\ a_{10} & a_{11} & a_{12} & \cdots \\ \vdots & \vdots & \vdots & \cdots \\ a_{i0} & a_{i1} & a_{i2} & \cdots \\ \vdots & \vdots & \vdots & \cdots \end{bmatrix}. \tag{41}$$

Equation (40) implies that each row of A sums to 0.

The generator matrix specifies both the parameters μ_i and the transition probabilities q_{ij}. The latter are given, according to (38), by

$$q_{ij} = \frac{a_{ij}}{\mu_i} = \frac{a_{ij}}{-a_{ii}}; \quad i \neq j = 0, 1, \ldots . \tag{42}$$

The transition probability functions of the Markov process are also determined by A. The relevant results are most conveniently presented in matrix form. Let $Q(t)$ be the matrix of transition probability functions $q_{ij}(t)$, where

$$q_{ij}(t) = P(X_t = j \mid X_0 = i); \quad i, j = 0, 1, \ldots . \tag{43}$$

This definition, which is equivalent to (37), allows us to interpret the ith row of $Q(t)$ as the distribution of the process state at time t, given that the initial state was i. In other words, $Q(t)$ describes the behaviour of the process over finite periods of time.

It can be shown (see exercise 3) that the following systems of differential equations are satisfied:

$$Q'(t) = AQ(t), \tag{44}$$

$$Q'(t) = Q(t)A. \tag{45}$$

These are known as the Kolmogorov backward and forward differential equations, respectively. Either (44) or (45) can be solved for $Q(t)$. A natural set of initial conditions is obtained by assuming that $Q(0)$ is the identity matrix:

$$q_{ij} = \begin{cases} 1 & \text{if } i = j \\ 0 & \text{if } i \neq j \end{cases}; \quad i, j = 0, 1, \dots . \tag{46}$$

If $Q(t)$ was an ordinary function and A was a constant, then the solution of (44) (or (45)) would be

$$Q(t) = e^{At}. \tag{47}$$

It turns out that this is a valid solution to our problem too, provided that (47) is interpreted as

$$Q(t) = \sum_{n=0}^{\infty} \frac{t^n}{n!} A^n; \quad t \geq 0. \tag{48}$$

It should be pointed out that the significance of this result is mainly theoretical, because of the difficulty in evaluating the right-hand side of (48). Problems associated with short-term or medium-term performance are usually tackled by applying an analytical or numerical solution method directly to the Kolmogorov differential equations. In a few simple cases, the solution can be found in closed form; in others, a numerical approach works quite well; by far the most common occurrence is that the problem is computationally intractable.

Our primary concern is with the long-term behaviour of a Markov process $X = \{X_t; t \geq 0\}$, with a state space S and generator matrix A. After the process has been running for a long time, the probability of observing it in state j should be independent of the initial state and should cease to vary with time (only the probability will become constant, not the state itself). Denote that limiting probability by p_j:

$$p_j = \lim_{t \to \infty} P(X_t = j \mid X_0 = i) = \lim_{t \to \infty} q_{ij}(t); \qquad j = 0, 1, \dots . \tag{49}$$

The analysis involved in establishing the existence of the limits (49), and in determining them, is very similar to the corresponding one for the case of a Markov chain. This is not surprising, when one considers the close relationship between a Markov process and a Markov chain. Indeed, if we ignore the times at which the process X moves from state to state, and simply number those transitions, then the resulting sequence of states, $\{X_n; n = 0, 1, \dots\}$, is a Markov chain. That chain is said to be 'embedded' in the process X. The embedded Markov chain has the same state space and the same initial state as the process X. Its one-step transition probabilities, q_{ij}, are given by (42).

A Markov process is defined to be irreducible, transient, recurrent null or recurrent non-null if its embedded Markov chain has those attributes. The question of periodicity does not arise in the case of the process because of the continuous time parameter.

There are two possibilities for the limiting probabilities of an irreducible Markov process.

(a) If the process is transient or recurrent null, then the limits p_j are equal to 0 for all j. When that is the case, we say that a steady-state distribution does not exist.

(b) If the process is recurrent non-null, then the limits p_j are positive for all j and sum up to 1. In that case, they constitute the steady-state distribution of the process.

It remains now to give a method for deciding whether (a) or (b) holds, and for determining the steady-state distribution when it exists. Such a method is provided by the following result:

Steady-state theorem for Markov processes. An irreducible Markov process X, with state space S and generator matrix $A = (a_{ij})$, $i, j \in S$, is recurrent non-null if, and only if, the system of equations

$$\sum_{i \in S} p_i a_{ij} = 0; \quad j \in S, \tag{50}$$

$$\sum_{j \in S} p_j = 1, \tag{51}$$

has a solution. That solution is then unique and is the steady-state distribution of the process X.

This steady-state theorem is very similar to the one given in the last section for Markov chains. As in that case, we shall content ourselves with giving an intuitive justification for the result.

We first rewrite equations (50) in the form

$$-a_{jj} p_j = \sum_{\substack{i \in S \\ i \neq j}} a_{ij} p_i; \quad j = 0, 1, \ldots,$$

which, after substitution of (40), becomes

$$\sum_{\substack{i \in S \\ i \neq j}} a_{ji} p_j = \sum_{\substack{i \in S \\ i \neq j}} a_{ij} p_i; \quad j = 0, 1, \ldots. \tag{52}$$

These are known as the 'balance equations' of the Markov process. If a steady-state distribution exists, then p_i $(i = 0, 1, \ldots)$ can be interpreted as the fraction of time that the process spends in state i. When it is in state i, it moves to state j at the rate of a_{ij}. Hence, $p_i a_{ij}$ is the average number of transitions that the process makes from state i to state j per unit time in the steady-state. That quantity will be referred to as the 'flow' from state i to state j.

Thus the right-hand side of (52) represents the total average number of transitions into state j per unit time in the steady-state (the total flow into state j). Similarly, the left-hand side represents the total average number of transitions out of state j per unit time in the steady-state (the total flow out of state j). Clearly, those two averages must be equal for the steady-state to exist.

This illustrates the necessity of (50) and (51). The sufficiency is demonstrated by arguing (as was done in the last section) that, if a solution for (50) and (51) exists, then the process can be neither transient, nor recurrent null.

It is very easy to write the balance equations (52) by looking at the state diagram of the Markov process. Remember that each arrow in that diagram – say from node i to node j – defines a flow, $p_i a_{ij}$. An imaginary cut is made in the diagram around node j (see figure 2.8). Then the total flow out of state j consists of all the arrows crossing the cut outwards, while the total flow into state j consists of all the arrows crossing the cut inwards.

Figure 2.8

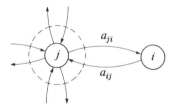

It is also possible to make cuts around groups of several nodes, and equate the total flows out with the total flows in. The resulting equations are not independent of the equations (52), but they may well be simpler and more convenient to solve.

As an application of the steady-state theorem, let us find the steady-state distributions of the Markov processes modelling the I/O buffer and the four-subscriber telephone network (examples 2 and 3). In both cases, the processes are irreducible and have finite state spaces, so we know beforehand that those distributions exist.

From the state diagram of the buffer process (figure 2.7a), by making a cut around each state in turn, we get the balance equations

$$\lambda p_0 = \mu(p_1 + p_2 + \cdots + p_M),$$
$$(\lambda + \mu)p_1 = \lambda p_0,$$
$$(\lambda + \mu)p_2 = \lambda p_1,$$
$$\vdots$$
$$\mu p_M = \lambda p_{M-1}.$$

To these we must add, of course, the normalising equation

$$p_0 + p_1 + \cdots + p_M = 1.$$

The solution of this system of equations is readily seen to be

$$p_j = \left(\frac{\lambda}{\lambda + \mu}\right)^j \frac{\mu}{\lambda + \mu}; \quad j = 0, 1, \ldots, M - 1; \quad p_M = \left(\frac{\lambda}{\lambda + \mu}\right)^M.$$

Having got the steady-state distribution, other performance measures such as the average number of messages in the buffer, the probability of losing a message due to finding the buffer full, etc., are easily obtained.

The balance equations for the telephone network process (see figure 2.7b) are

$$4\lambda p_0 = \mu p_1; \quad (\mu + \tfrac{2}{3}\lambda)p_1 = 4\lambda p_0 + 2\mu p_2; \quad \tfrac{2}{3}\lambda p_1 = 2\mu p_2.$$

Solving, say, the first and the third of these, together with the normalising equation, yields

$$p_0 = \left(1 + \frac{4\lambda}{\mu} + \frac{4\lambda^2}{3\mu^2}\right)^{-1}; \quad p_1 = \frac{4\lambda}{\mu}p_0; \quad p_2 = \frac{4\lambda^2}{3\mu^2}p_0.$$

The average number of conversations in progress is given by $p_1 + 2p_2$.

Exercises

1. Show that the initial conditions (46), together with relations (39), imply that the generators of a Markov process are equal to the derivatives of its transition probability functions at point 0: $a_{ij} = q'_{ij}(0)$, or $A = Q'(0)$.

2. Show that the transition probability functions satisfy a set of equations which, in matrix form, can be written as

$$Q(t + s) = Q(t)Q(s); \quad s, t \geqslant 0.$$

These are known as the 'Chapman–Kolmogorov equations'. They are the continuous analogue of (27) and can be derived in a similar way.

3. Demonstrate that (44) and (45) hold, by taking derivatives in the Chapman–Kolmogorov equations (see exercise 2) with respect to t and s, at $t = 0$ and $s = 0$ respectively. Use the result of exercise 1.

4. Draw a state diagram for the process K of example 1. Do equations (50) and (51) have a solution in this case? Is K transient, recurrent null or recurrent non-null?

5. In order to complete normally, a job has to go through M consecutive execution phases (e.g. compilation, linking, loading, etc.). The duration of phase i is exponentially distributed with parameter μ_i ($i = 1, 2, \ldots, M$). After completing phase i, the job starts phase $i + 1$ with probability α_i (for $i < M$), or is aborted with probability $1 - \alpha_i$ ($0 < \alpha_i < 1$). An aborted job, or one which completes phase M, is replaced immediately with a new job which starts phase 1.

Let X_t be the index of the phase which is being executed at time t. Show that $X = \{X_t; t \geq 0\}$ is an irreducible Markov process with state space $S = \{1, 2, \ldots, M\}$. Find the steady-state distribution of X. Hence obtain the average number of jobs that complete normally, and the average number that are aborted, per unit time.

Literature

A broad coverage and extra results on renewal processes can be found in Cox [7]. Our treatment of Poisson processes, Markov chains and Markov processes is similar in spirit to that of Cinlar [2]. The interested reader is directed to that book for additional information and for the rigorous proofs of the fundamental steady-state theorems. Among the many other books on stochastic processes which may be consulted are Doob [8] and Parzen [28].

3
Simple queueing models

The systems that we shall study here are composed of jobs, servers and queues, or service demands, service suppliers and waiting rooms, respectively. The framework provided by these 'queueing systems' is sufficiently general to encompass most computing, telecommunication and data processing activities (along with many others, as diverse as banking, manufacturing and transport). Depending on the context, the jobs may be computing tasks, input/output requests, telephone calls, data packets, etc. The servers may be processors, communication channels, software modules or user terminals. The queues may be primary or secondary storage areas, or external waiting lines.

This chapter is devoted to a special class of queueing systems, modelled by Markov processes of the so-called 'birth-and-death' type. The defining characteristic of these processes is that the only non-zero instantaneous transition rates out of state i ($i = 0, 1, \ldots$) are to states $i + 1$ and $i - 1$ (if $i > 0$). These transitions can be thought of as corresponding to a birth and a death, respectively; hence the name. In our models, the state of the process represents the number of jobs present in the system. Transitions from one state to another occur when jobs arrive into the system or when they depart from it. Then the above condition amounts to a requirement that jobs arrive and depart singly (rather than in batches).

When there is no danger of ambiguity, we shall refer to some well-known queueing systems by means of a shorthand notation devised by D. G. Kendall. This normally consists of three parts, describing the nature of the arrival process, the distribution of service times and the number of servers, respectively. For example, an M/D/2 system is one where jobs arrive in a Markov (Poisson) stream, service times are deterministic (i.e. of fixed length) and there are two servers.

3.1   The M/M/1 queue

Consider the simple queueing system illustrated in figure 3.1. Jobs arrive into the system demanding service, wait (if necessary) until they can

Figure 3.1

arrows ⟶ 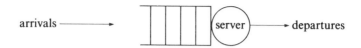 departures

be attended to, receive service and then depart. A single server is available at all times, which means that, whenever there are jobs in the system, one of them is being served. Jobs are selected for service in the order in which they arrive; once selected, they are served to completion. This scheduling strategy is referred to as first-come-first-served (FCFS), or first-in-first-out (FIFO). The queue may become arbitrarily long; jobs do not leave it before they are served.

Assume that the arrival process is Poisson, with rate λ, and that consecutive service times are independent random variables distributed exponentially with parameter μ. These assumptions ensure that both the arrivals and the services are memoryless (see section 2.2). More precisely, the following statements are true.

(i) The probability that an arrival occurs in a small interval of length h is equal to $\lambda h + o(h)$, regardless of the position of that interval and of anything that happened before it (the quantity $o(h)$ is negligible, compared with h).

(ii) If there is a service in progress at time t, that service will complete in the interval $(t, t + h)$ with probability $\mu h + o(h)$, regardless of the value of t and of anything that happened before t.

(iii) The probability that more than one arrivals and/or service completions occur in a small interval of length h is negligible; it is $o(h)$.

Let $N(t)$ be the number of jobs present in the system at time t (that number includes the job in service, if any). Our assumptions imply that $N = \{N(t); t \geqslant 0\}$ is a Markov process whose possible states are the non-negative integers $\{0, 1, \ldots\}$. Moreover, since transitions between states occur only at arrival and departure instants, the above statements (i), (ii) and (iii) indicate that the only non-zero instantaneous transition rates out of state i are to states $i + 1$ and $i - 1$ (if $i > 0$); those rates are λ and μ, respectively. Thus, N is a birth-and-death process, with a constant birth rate, λ, and a constant death rate, μ. The state diagram of that process is shown in figure 3.2.

Figure 3.2

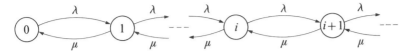

The model that we have described is known as the 'M/M/1 queue' (Markov arrivals, Markov services, single server). This is one of the few queueing models whose short-term behaviour can be analysed explicitly. In particular, there are closed form expressions for the probabilities, $p_{ij}(t)$, that the process will be in state j at time t, given that it was in state i at time 0 $(i, j = 0, 1, \ldots ; t \geqslant 0)$. However, neither the analysis nor the expressions are elementary.

We shall address a simpler problem, namely the long-term behaviour of the M/M/1 queue. Our first object is to determine the limiting probabilities,

$$p_j = \lim_{t \to \infty} p_{ij}(t); \quad i, j = 0, 1, \ldots ,$$

that the process is in state j, i.e. that there are j jobs in the system. Those limits exist, because the process N is irreducible (every state can be reached from every other state). However, they may all be equal to 0, in which case there would be no steady-state distribution.

Following the procedure suggested by the steady-state theorem for Markov processes (section 2.4), we write the system of balance equations for the probabilities p_j. This is best done by using the state diagram in figure 3.2: make an imaginary cut around each state in turn, and equate the flow out of the cut (average number of transitions out of that state per unit time), with the flow into the cut (average number of transitions into the state per unit time). The flow out of state 0 is λp_0, while the flow into it is μp_1. Hence

$$\lambda p_0 = \mu p_1 . \tag{1}$$

Similarly, the flow out of state j $(j > 0)$ is $(\lambda + \mu)p_j$ and the flow into it is $\lambda p_{j-1} + \mu p_{j+1}$:

$$(\lambda + \mu)p_j = \lambda p_{j-1} + \mu p_{j+1}; \quad j = 1, 2, \ldots . \tag{2}$$

An equivalent, but simpler system of equations can be obtained by adding together equations $0, 1, \ldots , j$. Most terms on both sides cancel, leaving only

$$\lambda p_j = \mu p_{j+1}; \quad j = 0, 1, \ldots . \tag{3}$$

These equations also represent balance of flow across cuts in the state diagram. Consider a cut enclosing the group of states $\{0, 1, \ldots , j\}$: only one arrow leaves it (from state j) and one arrow enters it (from state $j + 1$). The corresponding flows are precisely the left-hand and right-hand sides of (3). In general, any flow balance equation obtained from a cut enclosing an arbitrary subset of states is valid. Given a state diagram, one should look for the set of cuts that yields the most convenient set of balance equations.

Denote the ratio λ/μ by ρ. From equations (3), all probabilities p_j are easily expressed in terms of p_0:

$$p_j = \rho^j p_0; \quad j = 0, 1, \dots . \tag{4}$$

Now, the existence of a steady-state distribution hinges on whether it is possible to satisfy the normalising equation

$$\sum_{j=0}^{\infty} p_j = 1, \tag{5}$$

or, after substituting (4),

$$p_0 \sum_{j=0}^{\infty} \rho^j = 1. \tag{6}$$

Clearly, a solution to (6) exists only when the geometric series in the left-hand side converges, i.e. when $\rho < 1$. In that case (6) yields $p_0 = 1 - \rho$, and (4) gives the steady-state distribution of the number of jobs in the system as

$$p_j = \rho^j (1 - \rho); \quad j = 0, 1, \dots . \tag{7}$$

This is the modified geometric distribution that we encountered in section 1.4. It implies that observing the system in the steady-state is equivalent to performing a number of independent Bernoulli trials, such as tossing a coin. Suppose that each trial has two possible outcomes, called 'non-empty' and 'empty', which occur with probabilities ρ and $1 - \rho$, respectively. The experiment stops when the first 'empty' outcome occurs. Clearly, the number of 'non-empty' outcomes that occurred before that moment has the same distribution, (7), as the number of jobs in the M/M/1 system.

Note that the system parameters λ and μ do not appear in (7) except through their ratio, ρ. That important quantity has a simple physical meaning. Remember that λ is the average number of jobs that arrive into the system per unit time. The distribution of their required service times is exponential with parameter μ and so the average service time per job is $1/\mu$ (equation 2.(14)). Therefore, $\rho = \lambda/\mu$ is the average amount of required service time brought into the system per unit time. That ratio is referred to as the 'load', or 'traffic intensity' of the system.

The condition for existence of steady-state, $\rho < 1$, also has an intuitive interpretation. It says, in effect, that the demand for service must be less than the service capacity (in the case of a single server, the latter is 1 unit of service per unit time).

When $\rho \geqslant 1$, a steady-state distribution does not exist. We say then that the system is 'saturated', or 'overloaded'. The queue tends to grow and, in the long run, the probability of finding any finite number of jobs in the system is 0. In fact, one can distinguish two sub-cases of saturation. When $\rho = 1$, the Markov process N is recurrent null (see section 2.4): every state is visited infinitely many times, but at intervals which are infinitely long, on the average. When $\rho > 1$, the process is transient: every state is eventually visited for the last time; in the long run, the number of jobs in the system exceeds any finite number and does not fall below it again.

Let us now derive some performance measures for the $M/M/1$ system in the steady-state (assume that $\rho < 1$).

The utilisation of the server, U, is defined as the fraction of time that the server is busy. In other words, it is the probability that the system is not empty:

$$U = 1 - p_0 = \rho. \tag{8}$$

The same result can be obtained by arguing that, since each service lasts $1/\mu$ on the average, the server serves an average of μ jobs per unit time while it is busy. If it is busy for a fraction U of the time, then the overall departure rate is $U\mu$ jobs per unit time. But in the steady-state the departure rate must be equal to the arrival rate, hence $U\mu = \lambda$, or $U = \rho$. The advantage of this argument is that it does not rely on any assumptions about the distributions of interarrival and service times; only averages are involved here. Thus, in any single server system in the steady-state, the utilisation of the server is equal to the load.

Let L be the average number of jobs in the system. Using the definition of an average, L is obtained as

$$L = \sum_{j=1}^{\infty} jp_j = \rho(1-\rho) \sum_{j=1}^{\infty} j\rho^{j-1}$$

$$= \frac{\rho(1-\rho)}{(1-\rho)^2} = \frac{\rho}{(1-\rho)}. \tag{9}$$

This expression, which is plotted against ρ in figure 3.3, shows very clearly how the performance of the system deteriorates when the load increases: as $\rho \rightarrow 1$, the average number of jobs grows without bound. Moreover, the closer the system is to being saturated, the bigger the effect of increasing the load.

Figure 3.3

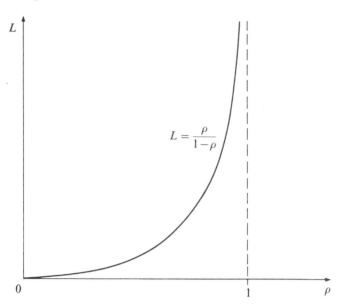

$$L = \frac{\rho}{1-\rho}$$

The average number of jobs in the queue, Q, is of course equal to the average number of jobs in the system minus the average number of jobs in service. Since there is a single server, the latter average is in fact the probability that the server is busy. Hence,

$$Q = L - U = \frac{\rho}{(1-\rho)} - \rho = \frac{\rho^2}{(1-\rho)}. \tag{10}$$

The performance indicator which is of most immediate concern to the users is the response time of a job, i.e. the time that a job spends waiting and receiving service. If the job finds a busy server, its waiting time consists of the remainder of the service in progress, plus the services of any other waiting jobs. Since service times are exponentially distributed, the remainder of a service has the same distribution as a whole service. Therefore, the response time of a job which finds, on arrival, j other jobs in the system ($j = 0, 1, \ldots$), has the same distribution as the sum of $j + 1$ service times. On the other hand, the random observer property of the Poisson process (section 2.2) implies that the number of jobs in the system seen by an arriving job has the same distribution, (7), as the number seen by a random observer.

Thus the average job response time, W, can be obtained from the relation

$$W = \sum_{j=0}^{\infty} (j+1) \frac{1}{\mu} p_j = (L+1) \frac{1}{\mu}. \tag{11}$$

Substitution of (9) yields

$$W = \frac{1}{\mu(1-\rho)} = \frac{1}{\mu - \lambda}. \tag{12}$$

The average waiting time of a job, W_Q (excluding the service), is equal to

$$W_Q = W - \frac{1}{\mu} = \frac{\rho}{\mu - \lambda}. \tag{13}$$

A realisation of a random response time can be constructed by means of the Bernoulli trials introduced after equation (7). Generate an exponentially distributed service time and then perform a trial. If the outcome is 'empty' stop; otherwise generate another service time, add it to the previous one and perform another trial; keep on, until the first 'empty' outcome occurs. Since this procedure generates $j + 1$ service times with probability $\rho^j(1 - \rho)$, it simulates precisely the response time of a job.

The above construction shows that the response time has the memoryless property. Indeed, both the service times and the Bernoulli trials are memoryless, so knowing that the experiment has been in progress for time x does not influence the probability that it will continue for time y more. That, in turn, implies that the response time is exponentially distributed (since no other continuous distribution is memoryless). The parameter of the distribution must be $\mu - \lambda$, to be consistent with the mean value (12).

We have thus established that the distribution function, $F(x)$, of a job's response time is given by

$$F(x) = 1 - e^{-(\mu - \lambda)x}; \quad x \geqslant 0. \tag{14}$$

Example

A single communication channel is used to send data items from several sources to a central computer. Assume that the stream of arrivals from each source is Poisson, with rate 2 items/second; moreover, the streams are independent of each other. All items are statistically identical, wait in a common queue and are transmitted one at a time. The transmission times can be assumed exponentially distributed, with mean 25 milliseconds.

It is required to determine the largest number of sources that can be connected to the channel. The following three criteria are suggested as possible grounds for reaching a decision.

(a) The channel must not be saturated.
(b) The steady-state average response time for an item must not exceed 100 milliseconds.
(c) In the steady-state, at least 95 % of all items should have response times not exceeding 100 milliseconds.

To answer these questions, we model a system with K sources connected to the channel by an M/M/1 queue with parameters $\lambda = 2K$ and $1/\mu = 0.025$. This is valid because, according to the superposition property of the Poisson process (section 2.2), the result of merging K independent Poisson processes is also Poisson.

The load generated by K sources is $\rho = 0.05K$. Hence, in order to satisfy (a), we must have $0.05K < 1$, or $K < 20$.

The average response time condition, (b), is expressed, using (12), as

$$\frac{1}{40 - 2K} \leqslant 0.1.$$

Solved for K, this yields $K \leqslant 15$.

If criterion (c) is adopted, then the probability that the response time of an item does not exceed 0.1 must be at least 0.95. In other words, $F(0.1) \geqslant 0.95$, or $1 - F(0.1) \leqslant 0.05$. After substitution of (14), this becomes

$$e^{-(40 - 2K)0.1} \leqslant 0.05,$$

or $K \leqslant 5$.

These numbers make quantitative the extent to which (c) is a stronger requirement than (b), which is stronger than (a).

Exercises

1. Use the steady-state distribution (7) to find the second moment, and hence the variance of the number of jobs in the system. Also, from (14), determine the variance of the response time. Verify that those variances are given by $\rho/(1 - \rho)^2$ and $1/(\mu - \lambda)^2$, respectively. Such results show that increasing the load leads not only to a worse average performance but to a more unpredictable one. Thus, in a heavily loaded system, a randomly observed queue size is likely to differ considerably from the average.

2. Suppose that an M/M/1 queue with parameters λ and μ is replaced by n independent M/M/1 queues, each with parameters λ/n and μ/n. For example, one fast processor is replaced by n slower ones (each having $(1/n)$th of its speed), and the single queue is replaced by n separate queues; jobs from the original arrival stream join each of these queues with probability $1/n$. Evaluate the effect of this change, using either the average response time or the average number of jobs in the system as a measure of performance.

3. Consider an M/M/1 queue where the jobs' willingness to join the queue is influenced by the size of the latter. More precisely, a job which finds j other jobs in the system joins the queue with probability $1/(j + 1)$ and departs immediately with probability $j/(j + 1)$ $(j = 0, 1, \ldots)$. The state diagram for this system is illustrated in figure 3.4.

Figure 3.4

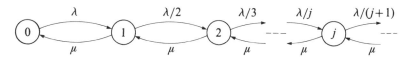

Write and solve the balance equations for the limiting probabilities p_j $(j = 0, 1, \ldots)$. Show that a steady-state distribution always exists. Find the steady-state utilisation of the server (the probability that it is busy), the throughput (average number of jobs departing per unit time), the average number of jobs in the system and the average response time for a job that decides to join.

4. Let D be a random interval between two consecutive departures from the M/M/1 queue in the steady-state. If, after the last departure, the queue was not empty, then D coincides with the service time of the next job. If the queue was empty, then D consists of the period until the next arrival, plus the service time of the new job. Arguing that the probability of a non-empty system just after a departure is the same as the utilisation of the server, i.e. U, show that D is exponentially distributed with parameter λ. (Hint: use the convolution formula 1.(68) to find the density function of the sum of an interarrival period and a service time.)

This result, which suggests that the departure process from the M/M/1 queue in the steady state is Poisson with rate λ, is a special case of Burke's theorem which will be discussed in the next section.

5. Show that the average number of jobs in the M/M/1 system, L, and the average response time there, W, are related as follows:

$$L = \lambda W.$$

This relation, to which we shall return many times in subsequent chapters, is valid under much more general assumptions.

6. Derive the result (14) directly from the expression

$$F(x) = \sum_{j=0}^{\infty} p_j F_{j+1}(x),$$

where p_j is given by (7) and $F_{j+1}(x)$ is the distribution function of the sum of $j + 1$ independent service times. That is the Erlang distribution function with parameters $j + 1$ and μ (see exercise 5 in section 2.2).

3.2   Multiple parallel servers

Let us generalise the M/M/1 model by assuming that service is provided by n identical and independent servers $(n > 1)$. All other assumptions are as before: the arrival process is Poisson with rate λ; there is a common queue where jobs wait in order of arrival; service times are distributed exponentially with parameter μ (at all servers). This is referred to as the M/M/n queue and is illustrated in figure 3.5.

Figure 3.5

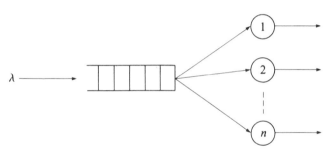

It is easily seen that $N = \{N(t); \ t \geqslant 0\}$, where $N(t)$ is the number of jobs in the system at time t, is an irreducible Markov process whose states are the non-negative integers $\{0, 1, \ldots\}$. The instantaneous transition rate from state j to state $j + 1$ is equal to the arrival rate, λ, for all $j = 0, 1, \ldots$. To determine the transition rate from state j to state $j - 1$, consider the two cases $j < n$ and $j \geqslant n$.

When $j < n$, all j jobs in the system are being served (by j of the servers). Since each service completes in a small interval of length h with probability $\mu h + o(h)$, the probability that there is a departure in that interval is equal to $j\mu h + o(h)$. Hence, the instantaneous transition rate from state j to state $j - 1$ is equal to $j\mu$, for $j = 1, 2, \ldots, n - 1$.

When $j \geqslant n$, there are n busy servers, i.e. n services in progress. The instantaneous transition rate from state j to state $j - 1$ is now independent of j and is equal to $n\mu$, for all $j = n, n + 1, \ldots$. Thus, the state diagram for the process N is as shown in figure 3.6.

Figure 3.6

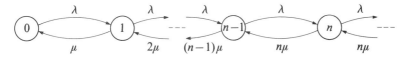

Denote the limiting probability of state j by p_j $(j = 0, 1, \ldots)$. These probabilities satisfy the following system of balance equations (the jth equation balances the flows out of, and into, a cut enclosing the set of states $\{0, 1, \ldots, j - 1\}$:

$$\lambda p_{j-1} = j\mu p_j; \quad j = 1, 2, \ldots, n - 1,$$

$$\lambda p_{j-1} = n\mu p_j; \quad j = n, n + 1, \ldots. \tag{15}$$

By successive elimination, all limiting probabilities can be expressed in terms of p_0:

$$p_j = \frac{\rho^j}{j!} p_0; \qquad j = 0, 1, \ldots, n - 1,$$

$$p_j = \frac{\rho^j}{n!\, n^{j-n}} p_0; \quad j = n, n + 1, \ldots. \tag{16}$$

Here, $\rho = \lambda/\mu$ is again the load on the system, i.e. the average amount of service demanded per unit time.

The service capacity of this system is n units of service per unit time (since there are n servers). Remembering that, for non-saturation, the load must be less than the service capacity, we can guess that the condition for existence of a steady-state distribution is $\rho < n$. This is indeed the case. In order that a steady-state distribution exists, there must be a solution (of the form (16)) to the balance equations, which also satisfies the normalising equation

$$\sum_{j=0}^{\infty} p_j = 1.$$

After substitution of (16), this becomes

$$\left[\sum_{j=0}^{n-1} \frac{\rho^j}{j!} + \frac{\rho^n}{n!} \sum_{i=0}^{\infty} \left(\frac{\rho}{n} \right)^i \right] p_0 = 1. \tag{17}$$

Equation (17) yields a non-zero value for p_0 only when $\rho < n$. Then we get

$$p_0 = \left[\sum_{j=0}^{n-1} \frac{\rho^j}{j!} + \frac{\rho^n}{(n-1)!\,(n-\rho)} \right]^{-1}. \tag{18}$$

When $\rho \geqslant n$, the system is saturated and a steady-state distribution does not exist. The long-run probability that there is any finite number of jobs in the system is 0.

Consider a non-saturated $M/M/n$ system in the steady-state, with a probability distribution given by (16) and (18). Let S be the average number of busy servers. That number can be obtained from the expression

$$S = \sum_{k=1}^{n-1} k p_k + n \sum_{k=n}^{\infty} p_k.$$

However, a more direct way of determining S, and one which has a more general validity, is to argue that since each server, while busy, serves an average of μ jobs per unit time, the overall departure rate is $S\mu$. In the steady-state, that must be equal to the arrival rate, λ. Hence,

$$S = \frac{\lambda}{\mu} = \rho. \tag{19}$$

As in the case of the $M/M/1$ system, this argument is independent of the exponential assumptions. The average fraction of busy servers, $U = S/n = \rho/n$, is referred to as the 'utilisation' of the set of servers.

The average number of jobs in the system, L, is equal to

$$L = \sum_{j=1}^{\infty} j p_j = \left[\sum_{j=1}^{n-1} \frac{\rho^j}{(j-1)!} + \frac{\rho^n(n^2 - n\rho + \rho)}{(n-1)!\,(n-\rho)^2} \right] p_0. \tag{20}$$

It is readily seen that, when the system approaches saturation ($\rho \to n$), the average number of jobs present approaches infinity.

Example

The steady-state performance of two systems – one $M/M/1$ with parameters λ and μ, and one $M/M/2$ with parameters λ and $\mu/2$ – is to be compared, using the average number of jobs, L, as a performance measure. This is a case of replacing one fast processor with two slow ones (each having exactly half the speed), but keeping a common queue. The conditions for existence of steady-state distributions, and the utilisations of the servers, U, are the same in both systems.

For the $M/M/1$ system we have, according to (8) and (9),

$$L_{M/M/1} = \frac{U}{1 - U},$$

where $U = \lambda/\mu$.

Applying the $M/M/2$ formulae (18) and (20), with $\rho = 2U$, yields

$$p_0 = \frac{1 - U}{1 + U}$$

and

$$L_{M/M/2} = \frac{2U}{(1 - U)(1 + U)}.$$

Bearing in mind that $U < 1$, we conclude that

$$L_{M/M/1} < L_{M/M/2},$$

i.e. the single-server system is more efficient than the two-server one. In general, it can be shown that if n increases and μ decreases, so that the product $n\mu$ remains constant (and hence U remains constant), the performance of the $M/M/n$ system deteriorates. The general formula (20) for L can be rewritten in terms of U as

$$L = nU + p_n \frac{U}{(1 - U)^2}, \tag{21}$$

where p_n is given by (16). The two terms in the right-hand side of (21) are, respectively, the average number of busy servers (or jobs in service) and the average queue size (jobs waiting). For fixed U, (21) is an increasing function of n, approaching nU when n is large. Intuitively, this deterioration in performance is explained by the fact that the processing power of an n-server system is utilised fully only when all servers are busy, i.e. when there are at least n jobs in the system. The more the processing power is fragmented into a large number of slow servers, the less efficiently it is being used. If, in addition to fragmenting the processing power, one fragments the queue (see exercise 2 in section 3.1), the performance becomes even worse.

Consider now the response time, t, of a job in the $M/M/n$ system. It is convenient to regard this as the sum of two independent random variables, $t = t_0 + t_1$, where t_0 is the waiting time (time spent in the queue) and t_1 is the service time. Of course, t_1 is known to be exponentially distributed, with mean $1/\mu$. The problem is thus to find the distribution function of t_0. Denote that distribution function by $G(x)$.

Let us introduce the probability, q, that a job has to wait, i.e. that it finds at least n jobs in the system. From (16) it follows that

$$q = P(t_0 > 0) = \sum_{j=n}^{\infty} p_j = \frac{\dfrac{\rho^n}{(n-1)!\,(n-\rho)}}{\displaystyle\sum_{j=0}^{n-1} \frac{\rho^j}{j!} + \frac{\rho^n}{(n-1)!\,(n-\rho)}}. \tag{22}$$

This is usually referred to as 'Erlang's delay formula'.

Conditioning upon not having, or having to wait, we can write, for $x \geqslant 0$,

$$G(x) = P(t_0 \leqslant x) = (1-q)P(t_0 \leqslant x \mid t_0 = 0) + qP(t_0 \leqslant x \mid t_0 > 0)$$
$$= 1 - q + qP(t_0 \leqslant x \mid t_0 > 0). \tag{23}$$

The next step is to remark that while all servers are busy, jobs depart from the system at intervals which are exponentially distributed with parameter $n\mu$ (superposition property of the Poisson stream). Therefore, as far as a waiting job is concerned, the queue behaves as if it is served by a single exponential server with parameter $n\mu$. Hence, the conditional waiting time of a job that has to wait in the $M/M/n$ system has the same distribution as the unconditional response time of a job in an $M/M/1$ system with parameters λ and $n\mu$. Substituting (14) into (23) we get

$$G(x) = 1 - q\mathrm{e}^{-(n\mu - \lambda)x}; \quad x \geqslant 0. \tag{24}$$

Note that this distribution function has a jump of size $1 - q$ at the origin.

The average waiting time, W_Q, is now easily obtained:

$$W_Q = \frac{q}{n\mu - \lambda}. \tag{25}$$

The distribution function of the response time is the convolution of (24) with the exponential distribution function with parameter μ. The average response time, W, is given by

$$W = W_Q + \frac{1}{\mu} = \frac{q}{n\mu - \lambda} + \frac{1}{\mu}. \tag{26}$$

Suppose now that the number of servers is infinite, so that there is never a queue and every incoming job starts service immediately. This is known as the $M/M/\infty$ system. In practice, of course, there are never infinitely many servers. Nevertheless, the $M/M/\infty$ system is not without interest. Later, when we look at network models, we shall encounter nodes where jobs do not compete for processors but are simply delayed for random periods of time, independently of each other. Such nodes can be thought of as containing infinitely many servers.

The steady-state probabilities, p_j, are obtained from (16):

$$p_j = \frac{\rho^j}{j!} p_0; \quad j = 0, 1, \ldots . \tag{27}$$

The normalising equation yields

$$1 = p_0 \sum_{j=0}^{\infty} \frac{\rho^j}{j!} = e^{\rho} p_0, \tag{28}$$

or $p_0 = e^{-\rho}$. Note that we always get a non-zero value for p_0, therefore a steady-state distribution always exists. This is only to be expected, since the service capacity is infinite and no load can saturate it.

Thus the steady-state number of jobs in the system, which is the same as the number of busy servers, has the Poisson distribution with parameter ρ:

$$p_j = \frac{\rho^j}{j!} e^{-\rho}; \quad j = 0, 1, \ldots . \tag{29}$$

The mean and the variance of that number are equal to ρ.

A rather remarkable feature of the above result is that it is in fact independent of the assumption that service times are exponentially distributed. If the $M/M/\infty$ system is replaced by an $M/G/\infty$ (Markov arrivals, general services, infinitely many servers) with the same load, ρ (arrival rate multiplied by the average service time), then (29) will continue to hold. Such 'insensitivity' with respect to the shape of the required service time distribution is associated with a particular type of system. One way (not quite accurate) of describing these systems is to say that every job that enters them starts receiving service immediately.

What can we say about the process of departures from a queueing system? We know that when the system is in the steady-state, the average number of departures per unit time is equal to the average number of arrivals per unit time. Does that equality extend to the distributions of arrivals and departures? In general, it does not. However, it turns out that when the system is $M/M/n$, or $M/M/\infty$, the departure process is statistically identical to the arrival process: it is Poisson (with rate λ). That result is known as 'Burke's theorem'. In fact, Burke's theorem makes a stronger statement: not only is the departure process Poisson, but its past history is independent of the current system state. This is a useful fact to bear in mind, especially in cases where the output from one system becomes the input, or part of the input, into another system.

An outline of a proof for Burke's theorem is given in exercises 2 and 3. If the number of servers is infinite, then the departure process is Poisson even when the service times have a general distribution.

Exercises

1. A computer system consists of five parallel processors, whose primary purpose is to serve user jobs. The latter arrive at the rate of 20 jobs per minute; each job takes an average of 9 seconds to execute. Whenever a processor has no user jobs to run (e.g. it completes an execution and finds an empty queue), it is assigned to secondary low-priority tasks, of which there is always an unlimited supply. The average execution time for a secondary task is 5 seconds. If a user job arrives and needs a processor which is running a secondary task, the latter is interrupted and eventually resumed when a processor is again available.

Find the average number of servers busy with user jobs and hence the service capacity available for the secondary tasks. From that, determine the throughput of the secondary tasks, i.e. the average number of them that are completed per minute.

2. Consider an M/M/n system in the steady-state, from the point of view of an observer looking backwards in time. Show that the instantaneous transition rates from state j to states $j + 1$ and $j - 1$ in reverse time are the same as the corresponding rates in forward time. For example,

$$P(\text{state } j + 1 \text{ at time } t - h \,|\, \text{state } j \text{ at time } t) = \lambda h + o(h);$$

$$j = 0, 1, \ldots .$$

(Hint: use the conditional probability formula $P(A\,|\,B)P(B) = P(B\,|\,A)P(A)$, together with the balance equations (15).) Thus the behaviour of the system with time reversed is indistinguishable from its behaviour in forward time. This is known as the 'reversibility property'; it holds for M/M/∞, as well as M/M/n queues.

3. Show that Burke's theorem is a consequence of the reversibility property, by arguing as follows: (i) the departure instants in forward time (transitions from state j to state $j - 1$) are precisely the arrival instants in reverse time (transitions from state j to state $j + 1$); (ii) the arrival process in reverse time is statistically identical to the arrival process in forward time, by the reversibility property; (iii) the arrival process in forward time is Poisson and its future is independent of the current system state; (iv) therefore, the departure process in forward time is Poisson and its past is independent of the current system state.

4. In an M/M/2 system with parameters λ and μ, one of the two processors behaves in an unusual fashion: it works only when the number of jobs in the system is odd. If a change from odd to even occurs while a service is in processs at that processor (e.g. a new job arrives, or one departs from the other processor), the service is interrupted, to be resumed from the point of interruption when the number becomes odd again.

Write the balance equations and find the condition for existence of steady-state. Explain why the apparently intuitive guess, $\rho < 1.5$, is wrong.

5. Consider a Markov process, $X = \{X_t; t \geqslant 0\}$, where the only non-zero instantaneous transition rates out of state i are to states $i + 1$ and $i - 1$ (if $i > 0$). These rates are denoted by

$$a_{i,i+1} = \lambda_i; \quad i = 0, 1, \ldots;$$
$$a_{i,i-1} = \mu_i; \quad i = 1, 2, \ldots.$$

This is the general formulation of a birth-and-death process. Write and solve the balance equations for the steady-state probabilities, p_i, and give the condition for existence of steady-state. Show that the M/M/1 queue and the M/M/n queue are special cases of the birth-and-death process.

3.3   Finite systems

There are several queuing models of the birth-and-death type, which give rise to processes with finite number of states. Bounds are imposed on the number of jobs that can be in the system at any one time, either by limiting the available storage space or by regulating the input mechanism. In such models, the question of existence of a steady-state distribution does not arise. As long as the process is irreducible (i.e. every state can be reached from every other state), steady-state exists, no matter how heavily the system is loaded. The balance and normalising equations can always be solved.

Perhaps the simplest example of a finite model is the M/M/1 queue with a limited waiting room. The system is as in section 3.1, but is constrained by having storage space for at most K jobs (including the one in service). Any job which finds, on arrival, K jobs in the system, is refused admittance and departs immediately. This is referred to as the M/M/1/K queue.

The possible process states are now $\{0, 1, \ldots, K\}$. The corresponding state diagram is obtained from the one in figure 3.2 by truncating it at state K (the only transitions out of, and into, state K are to, and from, state $K - 1$). The balance equations derived from that diagram are equivalent to equations (3), for $j = 0, 1, \ldots, K - 1$. They yield

$$p_j = \rho^j p_0; \quad j = 0, 1, \ldots, K, \tag{30}$$

where $\rho = \lambda/\mu$. This, together with the normalising equation

$$\sum_{j=0}^{K} p_j = 1, \tag{31}$$

implies that p_0 is equal to

$$p_0 = \frac{1 - \rho}{1 - \rho^{K+1}}. \tag{32}$$

Hence, the steady-state distribution of the $M/M/1/K$ queue is given by

$$p_j = \frac{\rho^j (1 - \rho)}{1 - \rho^{K+1}}; \quad j = 0, 1, \ldots, K. \tag{33}$$

Note the effect of the load on the shape of that distribution: When $\rho < 1$, the probabilities p_j decrease with j; the most likely state is the empty system (this is the case when steady-state exists for the unbounded $M/M/1$ queue). When $\rho = 1$, all states are equally likely (the right-hand side of (33) is indeterminate; however, an application of L'Hospital's rule shows that $p_j = 1/(K + 1)$, for all j). When $\rho > 1$, the probabilities p_j increase with j and the most likely state is the one where the queue is full. We also observe a predictable behaviour in the limit $K \to \infty$: (33) approaches the $M/M/1$ distribution (7) when $\rho < 1$, and it approaches 0, for all j, when $\rho \geqslant 1$.

The steady-state average number of jobs in the $M/M/1/K$ system is given by

$$L = \sum_{j=1}^{K} j p_j = \frac{\rho(1 - \rho^K)}{(1 - \rho)(1 - \rho^{K+1})} - \frac{K\rho^{K+1}}{1 - \rho^{K+1}}. \tag{34}$$

This expression has the value $K/2$ when $\rho = 1$.

Define the throughput, T, as the average number of jobs that are executed per unit time. This can be obtained from

$$T = (1 - p_0)\mu = (1 - p_K)\lambda \tag{35}$$

(jobs depart at rate μ while the server is busy; alternatively, they join at rate λ while the queue is not full).

The average response time, W, of a job that is admitted into the system can be found by remarking that such a job sees state j with probability $p_j/(1 - p_K)$. Therefore,

$$W = \sum_{j=0}^{K-1} (j + 1) \frac{1}{\mu} \frac{p_j}{1 - p_K}. \tag{36}$$

Multiplying and dividing the right-hand side by λ, and remembering that $\rho p_j = p_{j+1}$, equation (36) can be rewritten as

$$W = \frac{L}{T}. \tag{37}$$

The next system that we shall examine is one where a Poisson arrival stream of jobs (rate λ) is served by n parallel exponential servers (parameter μ), and there is space for at most n jobs. In other words, there is no queue; any job which finds all the servers busy departs immediately. This system, denoted by $M/M/n/n$, can be used to model a telephone exchange with n lines and no 'holding' facilities.

The number of jobs in the system (which is the same as the number of busy servers) is an irreducible Markov process whose possible states are $\{0, 1, \ldots, n\}$. The state diagram of that process is obtained from the one in figure 3.6 by truncating the latter at state n. The steady-state probabilities p_j satisfy equations (15), for $j = 1, 2, \ldots, n$, and can be expressed as

$$p_j = \frac{\rho^j}{j!} p_0; \quad j = 0, 1, \ldots, n, \tag{38}$$

where $\rho = \lambda/\mu$. The normalising equation then yields

$$p_0 = \left[\sum_{j=0}^{n} \frac{\rho^j}{j!} \right]^{-1}. \tag{39}$$

In the applications of the $M/M/n/n$ model to telephony, the chief performance measure of interest is the probability of losing a call, i.e. the probability of finding all lines busy, p_n. This is given by

$$p_n = \frac{\dfrac{\rho^n}{n!}}{\displaystyle\sum_{j=0}^{n} \frac{\rho^j}{j!}}. \tag{40}$$

Expression (40) is known as 'Erlang's loss formula'.

The system throughput, T, defined as the average number of jobs that are admitted (and completed) per unit time, is obtained from

$$T = \lambda(1 - p_n). \tag{41}$$

Example

A designer of a telephone exchange could be faced with the problem of determining the optimal number of lines for a given load. Assuming that the revenue from the exchange is proportional to the throughput and that its cost is proportional to the number of lines, the problem becomes one of maximising a non-linear objective function:

$$\max_{n \geqslant 1} \left[c_1 T - c_2 n \right],$$

where c_1 and c_2 are non-negative constants. This is equivalent to finding the minimum

$$\min_{n \geqslant 1} \left[\lambda c_1 p_n + c_2 n \right].$$

There is always an optimal value for n. Whether that value is 1, or is greater than 1, depends on the load ρ and on the ratio c_1/c_2.

Consider now a system where the jobs requiring service are generated by a finite number, K, of independent and statistically identical users. Each user, having submitted a job, waits until that job is completed, then 'thinks' for a while, then submits another job, and so on, *ad infinitum*. Thus, at any moment in time, j of the users are waiting for their jobs to be completed and $K - j$ users are thinking ($j = 0, 1, \ldots, K$).

Suppose, to start with, that there is a single server, serving jobs one at a time, in order of arrival. Service times are distributed exponentially with parameter μ. Think times are distributed exponentially with parameter λ. The system state, which can be defined as the number of jobs awaiting and/or receiving service, is then an irreducible Markov process with state space $\{0, 1, \ldots, K\}$.

One of the applications of this model has to do with machine availability: the users are K machines which break down from time to time; the server is a repairman, the jobs are the repair tasks generated by the breakdowns and the think times of a machine are the periods when it is operative. In that context, the term 'machine interference model' is sometimes employed.

We choose to think of the users as people sitting at terminals, and of the server as a processor. The jobs are program executions and the think times of a user are the delays between the completion of one program and the start of the next. With this application in mind, we shall talk about the 'terminal system model' (figure 3.7).

Figure 3.7

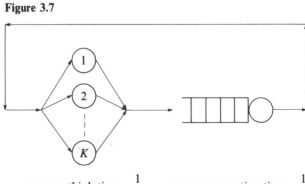

average think time $= \dfrac{1}{\lambda}$ average execution time $= \dfrac{1}{\mu}$

From state j, the process can move to state $j - 1$ (if $j > 0$ and the job in service completes), or to state $j + 1$ (if $j < K$ and one of the $K - j$ users in think state submits a new job). Since there is a single server, the instantaneous transition rate from state j to state $j - 1$ is the service completion rate, μ. The probability that, in a small interval of length h, a given user in think state will emerge from that state and submit a job, is $\lambda h + o(h)$. The probability that one of $K - j$ users will submit a job is $(K - j)\lambda h + o(h)$. Hence, the instantaneous transition rate from state j to state $j + 1$ is $(K - j)\lambda$. The state diagram of the process is illustrated in figure 3.8.

Figure 3.8

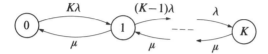

Denoting the steady-state probability of state j by p_j, we can write the following set of balance equations:

$$(K - j + 1)\lambda p_{j-1} = \mu p_j; \quad j = 1, 2, \ldots, K. \tag{42}$$

From these equations, all probabilities can be expressed in terms of p_0:

$$p_j = \frac{K!}{(K - j)!} \rho^j p_0; \quad j = 0, 1, \ldots, K, \tag{43}$$

where $\rho = \lambda/\mu$. The normalising equation yields the value of p_0:

$$p_0 = \left[\sum_{j=0}^{K} \frac{K!}{(K - j)!} \rho^j \right]^{-1}. \tag{44}$$

Let us derive some performance measures for the terminal system in the steady-state. The utilisation of the processor, U, is given by

$$U = 1 - p_0. \tag{45}$$

The system throughput, T, can be defined as the average number of jobs that are completed per unit time (this is, of course, equal to the average number of jobs that are submitted per unit time). Since jobs are completed at rate μ while the processor is busy, we have

$$T = U\mu = (1 - p_0)\mu. \tag{46}$$

The next important performance measure is the average response time, W (the average interval between the submission of a job and its completion). We have to find a new approach to this quantity, because here the arrivals do not form a Poisson stream, they do not behave like random observers and so the probability that a newly submitted job sees state j is not equal to p_j (for instance, the probability that such a job sees state K is 0).

Consider the instants of time when a given user submits jobs. The interval between one such instant and the next consists of the response time of the first job, followed by the ensuing think period. Hence, the average interval between two consecutive submissions is $W + (1/\lambda)$. This implies (see the arrival rate lemma in section 2.1) that the average number of jobs submitted by one user per unit time is $1/[W + (1/\lambda)]$. Since there are K users, the total throughput is K times that quantity.

Thus we have

$$T = \frac{K}{W + (1/\lambda)},\tag{47}$$

which can be solved for W:

$$W = \frac{K}{T} - \frac{1}{\lambda} = \frac{K}{(1 - p_0)\mu} - \frac{1}{\lambda}.\tag{48}$$

The above argument does not rely on the fact that think times and execution times are exponentially distributed. Hence, relations (47) and (48) hold in any terminal system with average think times $1/\lambda$ and average execution times $1/\mu$. The exponential assumptions were needed for the process to be Markov and for the validity of equations (43) and (44) (in fact, even those results are insensitive with respect to the distribution of think times; only the service times distribution can affect them).

The terminal system model can be generalised quite easily by assuming that service is provided by n identical parallel processors ($1 \leqslant n \leqslant K$). All other assumptions are as before. The resulting model is illustrated in figure 3.9.

Figure 3.9

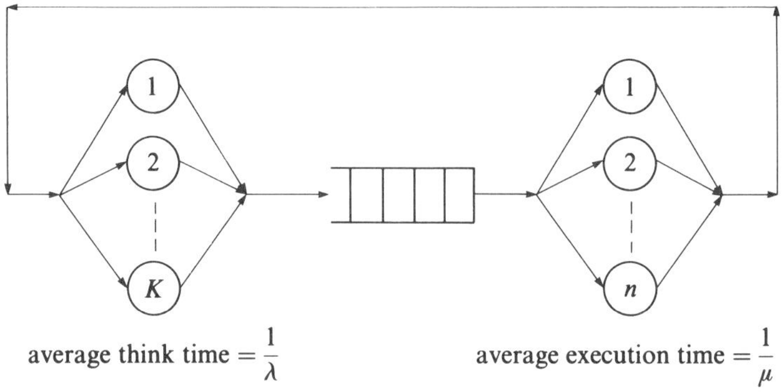

average think time $= \dfrac{1}{\lambda}$ average execution time $= \dfrac{1}{\mu}$

The instantaneous transition rate from state j to state $j - 1$ is now equal to $j\mu$ if $j < n$ (j jobs in service and each completes at rate μ), and it is equal to $n\mu$ if $j \geqslant n$ (n jobs in service). The new state diagram is shown in figure 3.10.

Figure 3.10

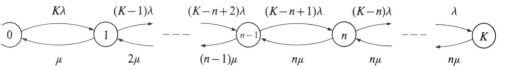

From this diagram, we obtain in the usual way a set of balance equations:

$$(K - j + 1)\lambda p_{j-1} = j\mu p_j; \quad j = 1, 2, \ldots, n - 1,$$
$$(K - j + 1)\lambda p_{j-1} = n\mu p_j; \quad j = n, n + 1, \ldots, K. \tag{49}$$

These equations, together with the normalising equation, yield

$$p_j = \frac{K!}{(K - j)! \, j!} \rho^j p_0; \quad j = 0, 1, \ldots, n - 1,$$

$$p_j = \frac{K!}{(K - j)! \, n! \, n^{j-n}} \rho^j p_0; \quad j = n, n + 1, \ldots, K, \tag{50}$$

and

$$p_0 = \left[\sum_{j=0}^{n-1} \frac{K!}{(K - j)! \, j!} \rho^j + \sum_{j=n}^{K} \frac{K!}{(K - j)! \, n! \, n^{j-n}} \rho^j \right]^{-1}. \tag{51}$$

The throughput, T, can be obtained either as the average number of job completions, or as the average number of job submissions, per unit time. The former approach requires the average number of busy servers, S:

$$S = \sum_{j=1}^{n-1} j p_j + n \sum_{j=n}^{K} p_j. \tag{52}$$

The expression for the throughput is then $T = S\mu$. Alternatively, we could find the average number of users in think state, M:

$$M = K - \sum_{j=1}^{K} j p_j. \tag{53}$$

Since each thinking user submits jobs at rate λ, the throughput is equal to $T = M\lambda$.

Having found T, we can go through exactly the same argument as the one that led to relation (48) and obtain the average response time for a job, W:

$$W = \frac{K}{T} - \frac{1}{\lambda}. \tag{54}$$

The special case when the number of processors is equal to the number of terminals ($n = K$) is the finite-user analogue of the $M/M/\infty$ system. In both cases, the users do not compete for the available resources and are, as a consequence, completely independent of each other. The steady-state distribution of the number of jobs in service is now given by (50) and (51), with $n = K$, which can be written as

$$p_j = \binom{K}{j} \left(\frac{\rho}{1 + \rho} \right)^j \left(\frac{1}{1 + \rho} \right)^{K-j}; \quad j = 0, 1, \dots, K. \tag{55}$$

This is the binomial distribution (section 1.4), with parameter $\rho/(1 + \rho)$. It implies that the number of jobs in service can be regarded as the number of successes in a set of K Bernoulli trials, where the probability of success at each trial is $\rho/(1 + \rho)$. That result can be explained intuitively by writing $\rho/(1 + \rho) = (1/\mu)/[(1/\lambda) + (1/\mu)]$. The probability of success can then be interpreted as the fraction of time that a given user spends waiting for a job to complete. The probability of failure, $1/(1 + \rho) = (1/\lambda)/[(1/\lambda) + (1/\mu)]$, is the fraction of time that a given user spends in the think state. This interpretation, as well as (55), is valid for any distribution of think times and execution times.

The average number of busy servers is given by $S = K\rho/(1 + \rho)$. The throughput is $T = S\mu$ and the average response time is, of course, equal to the average execution time: $W = 1/\mu$.

Referring again to the machine availability application, the present case represents K machines which break down and are repaired independently of each other (there is a repairman for each machine). Expression (55) gives the distribution of the broken down machines or, if λ and μ are exchanged, the distribution of the operative machines.

Exercises

1. A fixed number, K, of jobs circulate among two service nodes, each with its own queue and a processor. After completing service at node 1, a job goes to node 2 with probability a and leaves the system with probability $1 - a$; in the latter case, the job is immediately replaced by a new job at node 1. After completing service at node 2, jobs return to node 1. Service times at nodes 1 and 2 are exponentially distributed, with parameters μ and λ, respectively (figure 3.11).

Figure 3.11

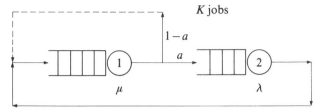

Let p_j be the steady-state probability that there are j jobs at node 1 (waiting and/or in service); $j = 0, 1, \ldots, K$. Show, by writing and solving the balance equations, that p_j is equal to the probability that there are j jobs in an M/M/1/K system with arrival rate λ and service time parameter $a\mu$. Find the throughput of the present system, i.e. the average number of departures per unit time.

2. In a terminal system with K users and a single processor, a charge of c per unit of processor time is levied. What is the average revenue per unit time? One fine day, the management decides to double the charge. As a result, half of the users disconnect their terminals and transfer their custom elsewhere (assume that K is even). Under what conditions, if any, is that move profitable?

3. Modify the terminal system model with n processors $(n \leqslant K)$, by assuming that there is no room for a queue: any job which is submitted when all the processors are busy is discarded and the corresponding user enters a new think period. Solve the balance equations and find the distribution of the number of jobs in service (this is known as the 'Engset distribution'). Hence derive expressions for the throughput (average number of successfully completed jobs per unit time) and the average number of discarded jobs per unit time.

4. Generalise the model from example 3 in section 2.4 by considering a telephone network with K subscribers (assume that K is even). Find the steady-state probability, p_j, that there are j conversations in progress $(j = 0, 1, \ldots, K/2)$, the average number of calls lost per unit time and the average revenue per unit time.

Literature

Queuing theory emerged as a separate subject through the work of A. K. Erlang (circa 1920) and other pioneers such as T. Engset, C. Palm, F.

Pollaczek and A. Y. Khinchin. Although the original motivation for the research came from various problems in telephony, it was soon realised that the field of applications for the theory was very wide. The advent of computers enlarged that field even further and added considerably to its importance.

There is an extensive body of literature on both the theory and the applications of queueing models. A number of books cover different parts of the material in the present chapter. The interested reader could consult, for instance, T. L. Saaty [29] or volume 1 of L. Kleinrock [20]. Both these texts contain many other references.

4

General service times and different scheduling strategies

The models that we shall examine here differ in some important respects from the birth-and-death ones introduced in the last chapter. The general framework will be that of a single-server queueing system with Poisson arrivals. However, it will no longer be assumed that service times are exponentially distributed. The first object of interest in this connection will be the $M/G/1$ queue (Markov arrivals, general service times) with FIFO (first-in-first-out) scheduling of jobs. Some performance measures for that system can be obtained by a direct and intuitive approach; for others, an analytical device that restores the Markov nature of the process has to be employed.

The assumption that all jobs are statistically identical will also be relaxed. The job population may be split into different classes, with different arrival and required service time characteristics. Finally, the order in which jobs are selected for service need not be FIFO. Other scheduling strategies commonly used in computer systems will be considered and evaluated. Some problems of optimal scheduling will be addressed.

Before embarking on the study of particular models, we shall introduce a very important and useful relation between the average number of jobs and the average response time in a queuing system in the steady-state. That relation is named after J. D. C. Little, who first proved it in the general case.

4.1   Little's result

Consider an arbitrary queueing system, Ξ. The internal structure of Ξ is of no importance at all; it can be regarded simply as a place where jobs arrive, remain for some time, and then depart. The only assumption that we shall make is that Ξ is in the steady-state. In other words, there are fixed probabilities of observing Ξ in its various possible states and those probabilities sum up to 1.

Denote by L the average number of jobs present in Ξ and by W the average time that jobs spend in Ξ (the average response time). Let λ be the arrival rate, i.e. the average number of jobs coming into Ξ per unit time. Then the following relation holds:

$$L = \lambda W. \tag{1}$$

This is Little's result. Several proofs for it have been proposed, each of them simpler and more elegant than the previous ones. The shortest of those proofs is outlined in exercise 1. Here we shall be content with providing a convincing intuitive justification for the validity of the relation.

Suppose that a charge of 1 is levied on each job in the system, for each unit of time that it spends there. Then, if the money is collected continuously, the total average revenue per unit time is equal to the average number of jobs in Ξ, which is L. On the other hand, the entire contribution from one job is equal to the time the job spends in Ξ; its average value is W. That entire contribution can be collected at the instant of the job's arrival (or departure). Since an average of λ jobs arrive into (and leave) Ξ per unit time, this alternative collection method yields an average revenue of λW per unit time. However, it is clear that, in the long run, the revenues produced by the two methods should be the same. Hence, we must have $L = \lambda W$.

The above argument is obviously very general. It uses only averages and the assumption of stationarity. Nothing is said about the distributions of interarrival and service times, number of servers, scheduling strategies or dependencies between jobs. We are completely free to define both Ξ and the jobs in it as we wish. That freedom can be exploited to considerable advantage.

The most immediate benefit of Little's result is that, if the arrival rate λ is known, it is enough to determine somehow one of the performance measures L and W; the other one is then given by relation (1). For instance, the average response time in the $M/M/n$ system can be obtained from $W = L/\lambda$, with L given by 3.(20), rather than from expression 3.(26).

Sometimes, the addition of another relation between L and W allows both those quantities to be determined simultaneously. In the case of the $M/M/1$ queue, for example, one can argue as follows. When a job arrives in the system, it sees an average of L jobs there. Each of those jobs (including the one in service, if any) will delay it for an average of $1/\mu$, as will the job's own service. Hence, the average response time can be expressed as

$$W = (L + 1)\frac{1}{\mu}. \tag{2}$$

Little's result can now be used to eliminate either L or W, yielding an equation which determines the other unknown. Substituting λW for L, we get

$$W = (\lambda W + 1)\frac{1}{\mu}, \tag{3}$$

or $W = 1/(\mu - \lambda)$. Having found W, the value of L is obtained from $L = \lambda W = \rho/(1 - \rho)$. This derivation produces the right answers (see equations 3.(9) and 3.(12)), without needing the steady-state probabilities p_j.

The technique of determining an unknown quantity by expressing it in terms of itself has many applications in modelling. What we have in (3) is a simple example of a 'fixed-point equation', i.e. a relation of the form

$$X = f(X), \tag{4}$$

where X is an unknown scalar or vector, and f is a given function or operator. The solution of a fixed-point equation may be elementary, as in the case of (3), or it may require some iterative numerical procedure. We shall encounter such equations on a number of occasions in this and subsequent chapters. Many of them will be obtained with the aid of Little's result.

Exercises

1. Denote by $F(x)$ the probability distribution function of the time that a job spends in the system Ξ. Let t be an arbitrary observation point in the steady-state. The number of jobs that are in Ξ at time t comprises those whose arrival is before t and whose departure is after t. Consider the contribution to that number made by an infinitesimal interval in the past, $(t - x, t - x + \mathrm{d}x)$. Argue that the average contribution is $\lambda[1 - F(x)]\mathrm{d}x$, since an average of $\lambda\mathrm{d}x$ jobs arrive during such an interval and each of them is still in the system at time t with probability $1 - F(x)$. Write an expression for L by integrating those average contributions over all x from 0 to ∞. Show that that expression is equal to λW, thus establishing Little's result (see also exercise 6 in section 1.3).

2. The demand in a multiprogrammed computer system is so heavy that the management decides to operate an admission policy which restricts the degree of multiprogramming to 60. This means that 60 jobs share the computing resources at all times and, as soon as one of them completes, a new job is admitted from a never-empty pool of waiting jobs. The resulting throughput turns out to be 5 jobs per second, on the average. Does this fact imply anything about the average response time of a job (the time between its admission and completion) and if so, what?

4.2   The M/G/1 queue

When constructing a model of a single-server system, it is usually easier to justify an assumption of Poisson arrivals than one of exponentially distributed service times. If the arrival stream is formed by merging together requests from a large number of independent sources, then a limit theorem can be invoked to assert that the stream is approximately Poisson (see section 2.2). A similar claim for the exponentiality of the service time distribution would rest on a much weaker foundation. Indeed, one can readily think of systems where the service times are patently not exponentially distributed (e.g. a communication channel transmitting data packets of fixed length).

It is thus desirable to study a single-server model where jobs arrive according to a Poisson process (rate λ) and service times are independent and identically distributed random variables with some general distribution function, $F(x)$. Assume that there is no bound on the number of jobs that may wait for service and that the scheduling strategy is FIFO. This model is referred to as the 'M/G/1 queue'.

The greater generality of the M/G/1 model is, of course, bought at a price. When the service time distribution does not have the memoryless property, the number of jobs in the system at time t, $N(t)$, is not a Markov process. The behaviour of that number after time t depends not only on its value at t, but also on when the current service will complete. That moment, in turn, depends on when the current service started, i.e. on something that occurred before time t. Consequently, the steady-state distribution of the number of jobs in the system cannot be obtained by the methods that we have used so far.

Some performance measures can be derived quite simply. Denote, as before, the average required service time by $1/\mu$. The load (the average amount of work arriving into the system per unit time) is $\rho = \lambda/\mu$. Since the single server can do at most one unit of work per unit time, the condition for non-saturation is $\rho < 1$. We shall assume that that condition is satisfied, and that the queue is in steady-state.

The utilisation of the server, U, defined as the probability that the server is busy, or as the average number of jobs in service, is equal to the load. The argument which uses the equality of the departure rate, $U\mu$, and the arrival rate, λ, to deduce $U = \rho$, applies here too.

Let L and L_Q be the average numbers of jobs in the system and in the queue, respectively (the latter number excludes the job in service, if any). Similarly, let W and W_Q be the average response time and queuing time (the latter excludes the service), respectively. Applying Little's result to the entire system (queue plus server) and to the queue only, we get the relations

$$L = \lambda W; \quad L_Q = \lambda W_Q. \tag{5}$$

In addition, the average response time is obviously equal to the average queuing time plus the average service time:

$$W = W_Q + \frac{1}{\mu}. \tag{6}$$

We thus have three equations involving the four unknown quantities L, L_Q, W and W_Q. One more independent equation would enable us to determine all four unknowns. To find that extra equation, consider the components of a job's queuing time. A newly arriving job has to wait for the current service, if any, to complete, and also for all jobs in the queue to receive their services, before it can start its own service (figure 4.1).

Figure 4.1

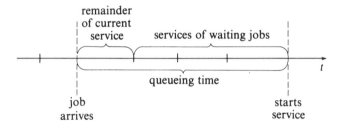

The probability that there is a service in progress when a new job arrives into the system is equal to the server utilisation, which is ρ (once again, we use the random observer property of the Poisson process). If there is a service in progress, then its remaining duration is the quantity that was referred to as 'residual life' in section 2.1. Applying formula 2.(11), we can write

$$\text{Average remaining service time} = \frac{\mu M_2}{2}, \tag{7}$$

where M_2 is the second moment of the service time:

$$M_2 = \int_0^\infty x^2 \mathrm{d}F(x).$$

The average number of jobs that the new arrival sees waiting in the queue is L_Q. Each of those jobs will take an average of $1/\mu$ to serve. Hence, the average queuing time can be expressed as

$$W_Q = \rho \frac{\mu M_2}{2} + \frac{1}{\mu} L_Q = \frac{\lambda M_2}{2} + \frac{1}{\mu} L_Q. \tag{8}$$

This is the desired fourth equation. Eliminating L_Q with the aid of (5), we get a fixed-point equation for W_Q:

$$W_Q = \frac{\lambda M_2}{2} + \rho W_Q. \tag{9}$$

This yields

$$W_Q = \frac{\lambda M_2}{2(1 - \rho)}. \tag{10}$$

The other averages are now easily determined from (5) and (6). For instance, the average number of jobs in the system is given by

$$L = \rho + \frac{\lambda^2 M_2}{2(1 - \rho)}. \tag{11}$$

This is known as the 'Pollaczek–Khinchin formula'. It is usually presented in a slightly different form, using a characteristic called the 'squared coefficient of variation' instead of the second moment of the service time. For a given random variable X, the squared coefficient of variation, C^2, is defined as the ratio of the variance to the square of the mean:

$$C^2 = \frac{\text{Var}(X)}{[E(X)]^2} = \frac{E(X^2) - [E(X)]^2}{[E(X)]^2} = \frac{E(X^2)}{[E(X)]^2} - 1. \tag{12}$$

Conversely, the second moment of the random variable can be expressed in terms of its squared coefficient of variation:

$$E(X^2) = [E(X)]^2(1 + C^2). \tag{13}$$

Using this expression, the Pollaczek–Khinchin formula can be rewritten as

$$L = \rho + \frac{\rho^2(1 + C^2)}{2(1 - \rho)}. \tag{14}$$

We observe that, just as in the case of the $M/M/1$ queue, the average number of jobs in the system increases without bound when the load approaches 1. The distinctive feature of the present model is the appearance of the second moment of the service time (or its coefficient of variation) as a factor which influences performance. Even a lightly loaded $M/G/1$ queue can perform very poorly when the variability of the demand is high. For a fixed value of the load, ρ, the best performance is achieved when $C^2 = 0$, i.e.

when the variance of the service times is 0. In other words, the ideal pattern of demand is the one where all jobs require exactly the same amount of service.

Example

Let us apply the M/G/1 model to the performance evaluation of a fixed-head peripheral storage device such as a paging drum. The device contains a number of circular tracks, where information is recorded in the form of fixed-size records. In this context, 'a job' is a request to read or write one record on a particular track. Each track is equipped with its own read/write head and is thus independent of all the other tracks. This makes it possible to treat an isolated track as a single-server queuing system.

Assume that the read/write requests for the given track arrive according to a Poisson stream with rate λ, and that they queue and are served in FIFO order. There are n records on the circumference of the track and each request is equally likely to be addressed at any of them. The service of a request consists of waiting until the corresponding record comes under the read/write head, and then reading or writing it. The operation of the drum is illustrated in figure 4.2.

Figure 4.2

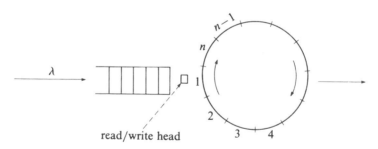

read/write head

The speed at which the drum revolves is R (revolutions per unit time). Hence, the time for one record to pass under the read/write head is $1/(nR)$. Suppose that, at the completion of a service, the ith record has just passed under the head. If the target of the next request is record $i + 1$, then its service time will be $1/(nR)$; it will be $2/(nR)$ if the target is record $i + 2; \ldots; n/(nR) = 1/R$ if the target is again record i. Since all those occurrences are equally likely, we shall define the service time of a request, S, as a random variable which takes the value $j/(nR)$ $(j = 1, 2, \ldots, n)$ with probability $1/n$. In fact, that definition is a slight misrepresentation of reality: a request arriving into an empty system does not have to find the head at the end of a record; its service time can take any value in the interval $(1/(nR), 1/R)$. However, when n is large, and/or when the drum is heavily loaded, the error of the approximation is negligible.

The first and second moments of the service time, and its squared coefficient of variation, are given by

$$E(S) = \sum_{j=1}^{n} \frac{j}{nR}\frac{1}{n} = \frac{n+1}{2nR},$$

$$E(S^2) = \sum_{j=1}^{n} \left(\frac{j}{nR}\right)^2 \frac{1}{n} = \frac{(n+1)(2n+1)}{6n^2 R^2},$$

$$1 + C^2 = \frac{E(S^2)}{[E(S)]^2} = \frac{2(2n+1)}{3(n+1)}.$$

The load is thus equal to $\rho = \lambda E(S) = \lambda(n+1)/(2nR)$. In order that the drum does not saturate, we must have $\rho < 1$, or $\lambda < 2nR/(n+1)$. This condition restricts the acceptable average demand rate to less than two requests per revolution.

The steady-state average number of tasks present, L, and the average response time, W, are obtained from the Pollaczek–Khinchin formula and Little's result. For instance, when n is large, the average response time is approximately equal to

$$W \approx \frac{1}{2R} + \frac{\lambda}{3R(2R - \lambda)}.$$

The first term in this expression is the average service time and the second is the average queuing time.

Suppose now that we are interested in performance measures of the $M/G/1$ queue that involve distributions, rather than just averages. We may wish to determine the steady-state probabilities, p_j, that there are j jobs in the system ($j = 0, 1, \ldots$). Or the distribution function, $H(x)$, of the response time. The approach that we have used before – namely writing and solving a set of balance equations for p_j – is not directly applicable here because the process involved is not Markov. It turns out, however, that the Markov property holds at selected moments in time. By considering the system state only at those moments, one can define a Markov chain whose steady-state distribution can be obtained by standard methods.

The reader who is not interested in distributions, or is put off by the analysis, may skip the rest of this section.

Let t_n be the instant when the nth departure from the system occurs ($n = 1, 2, \ldots$), and let X_n be the number of jobs in the system at time t_n^+, i.e. just after the nth departure. Clearly, if the value of X_n is known, then the value of X_{n+1} does not depend on anything that happened before t_n. Indeed, the arrivals following t_n are not influenced by past events because the arrival process is Poisson; the departures following t_n are not influenced by past events because there is no service in progress at t_n. All this implies that $X = \{X_n; n = 1, 2, \ldots\}$ is a Markov chain.

In the steady-state, the distribution of the number of jobs left behind by a departure is the same as the distribution of the number of jobs seen by an arrival. This is because the fraction of departures that leave the system in state j is equal to the fraction of arrivals that see it in state j (see exercise 3). On the other hand, the distribution of the number of jobs seen by an arrival is the same as the distribution of the number of jobs seen by a random observer, according to the random observer property of the Poisson stream. Therefore, if we find the limiting distribution of the Markov chain X,

$$p_j = \lim_{n \to \infty} P(X_n = j); \quad j = 0, 1, \ldots , \tag{15}$$

we shall also have the steady-state distribution of the number of jobs seen in the system by a random observer.

According to the steady-state theorem for Markov chains (section 2.3), the probabilities p_j satisfy the set of balance equations

$$p_j = \sum_{i=0}^{\infty} p_i q_{ij}; \quad j = 0, 1, \ldots , \tag{16}$$

where q_{ij} is the one-step transition probability from state i to state j. We already know that these equations will have a solution which satisfies the normalising equation if $\rho < 1$. Moreover, since the utilisation of the server is equal to ρ, we must have $p_0 = 1 - \rho$.

Suppose that $X_n = i$, for $i \geqslant 1$, and consider the possible values of X_{n+1}. At time t_n, one of the i jobs in the system starts its service; that job departs at t_{n+1}. If no new jobs arrive during the service time, X_{n+1} will be equal to $i - 1$; if one new job arrives, then $X_{n+1} = i$; if two jobs arrive, then $X_{n+1} = i + 1$, etc. Denoting by r_k the probability that exactly k jobs arrive into the system during a service time, we can write

$$q_{ij} = P(X_{n+1} = j \mid X_n = i) = r_{j-i+1};$$
$$i = 1, 2, \ldots; \; j = i - 1, i, \ldots . \tag{17}$$

If $X_n = 0$, then t_n is the start of an idle period. Eventually, a job arrives into the system and starts service; that job departs at t_{n+1}. If j jobs arrive during the service, then X_{n+1} will be equal to j. In other words,

$$q_{0j} = P(X_{n+1} = j \mid X_n = 0) = r_j; \quad j = 0, 1, \ldots . \tag{18}$$

Substituting (18) and (17) into (16), we get the set of equations

$$p_j = p_0 r_j + \sum_{i=1}^{j+1} p_i r_{j-i+1}; \quad j = 0, 1, \ldots . \tag{19}$$

The probabilities r_k are calculated by remarking that the number of arrivals during an interval of length x has the Poisson distribution with parameter λx. The probability that a service time is of length x is $f(x)\mathrm{d}x$, where $f(x)$ is the service time density function (see section 1.2; this should be replaced by $\mathrm{d}F(x)$ if the density function does not exist). Hence,

$$r_k = \int_0^\infty \frac{(\lambda x)^k}{k!} \, \mathrm{e}^{-\lambda x} f(x) \mathrm{d}x; \quad k = 0, 1, \ldots . \tag{20}$$

Having got the values of r_k, and knowing that $p_0 = 1 - \rho$, equations (19) can be solved by successive elimination. For instance, the equation for $j = 0$ yields $p_1 = (1 - \rho)(1 - r_0)/r_0$. A numerical solution can be obtained in this fashion quite easily. However, the general expression for p_j is too unwieldy to be written down in closed form. A much more elegant way of dealing with the set (19) is to transform it into a single equation by introducing the generating functions, $p(z)$ and $r(z)$, of the distributions p_j and r_k, respectively:

$$p(z) = \sum_{j=0}^\infty p_j z^j; \quad r(z) = \sum_{k=0}^\infty r_k z^k. \tag{21}$$

Multiplying the jth equation in (19) by z^j and summing over j from 0 to ∞ we get

$$p(z) = p_0 r(z) + \sum_{j=0}^\infty \sum_{i=1}^{j+1} p_i r_{j-i+1} z^j. \tag{22}$$

If, in the second term in the right-hand side of (22), z^j is replaced by $z^i z^{j-i+1} z^{-1}$, the order of summation is changed and the two sums are separated, that equation becomes

$$p(z) = p_0 r(z) + \frac{1}{z} r(z) [p(z) - p_0]. \tag{23}$$

The unknown generating function $p(z)$ can now be expressed in terms of $r(z)$:

$$p(z) = \frac{p_0 (1 - z) r(z)}{r(z) - z}. \tag{24}$$

It remains to give an expression for $r(z)$. This comes from (20), after multiplying by z^k and summing:

$$r(z) = \int_0^\infty \sum_{k=0}^\infty \frac{(\lambda xz)^k}{k!} e^{-\lambda x} f(x) \mathrm{d}x$$

$$= \int_0^\infty e^{-\lambda x(1-z)} f(x) \mathrm{d}x. \tag{25}$$

Remembering the definition of a Laplace transform (section 1.5), we notice that the right-hand side of (25) is precisely the Laplace transform of the service time density function, taken at the point $\lambda(1-z)$. Thus we have

$$r(z) = f^*(\lambda - \lambda z). \tag{26}$$

This relation between the Laplace transform associated with a random interval of time (in this case a service time), and the generating function of the number of arrivals during that interval, is of interest in its own right. We shall refer to it again shortly.

Substituting (26) into (24), we obtain the distribution of the number of jobs in the $M/G/1$ system in an explicit, if not very intuitively appealing form:

$$p(z) = \frac{(1-\rho)(1-z)f^*(\lambda - \lambda z)}{f^*(\lambda - \lambda z) - z}. \tag{27}$$

Formula (27) can be used to determine the moments of the number of jobs in the system. For instance, the average, L, is given by $p'(1)$. In working out the derivative for $z = 1$, one has to resolve an indeterminacy in the right-hand side of (27) (since $f^*(0) = 1$). This can be done by means of L'Hospital's rule. What emerges then is the Pollaczek–Khinchin formula, (11). Indeed, that is how that formula was originally derived.

The distribution of the response time is obtained from the following simple observation: The jobs that are in the system at the moment of a job's departure are precisely the ones that arrived during that job's response time. Therefore, we can write a relation similar to (26), between the generating function of the number of jobs in the system (at departure instants or at random points) and the Laplace transform of the density function of the response time. Denoting the latter Laplace transform by $h^*(s)$, we get

$$p(z) = h^*(\lambda - \lambda z). \tag{28}$$

This is transformed into an expression for $h^*(s)$ by making a change of variables, $s = \lambda - \lambda z$, and substituting (27):

$$h^*(s) = p\left(\frac{\lambda - s}{\lambda}\right) = \frac{(1-\rho)sf^*(s)}{\lambda f^*(s) - \lambda + s}. \tag{29}$$

By taking derivatives in (29) at point $s = 0$, one can determine the moments of the response time. The average is, of course, already specified by the Pollaczek–Khinchin formula and Little's result.

Exercises

1. Two applicants are being considered for a vacancy as a bank teller. By performing empirical tests, it is established that the service times of the first applicant are distributed exponentially with mean 0.9 minutes, while those of the second are uniformly distributed between 0.8 and 1.2 minutes. It is known that customers arrive at that particular teller's window in a Poisson stream, at the rate of 50 per hour.

Can both candidates cope with the load? If so, which one should be employed in order to minimise the average queue size?

2. Alter the paging drum model from the example in this section by assuming that there is a separate queue for the requests directed at record i $(i = 1, 2, \ldots, n)$. As record i passes under the read/write head, the first request in that queue, if any, is serviced.

Ignoring the slightly different behaviour of requests that arrive into empty queues, show that each of the n queues can be treated as an independent $M/G/1$ queue with arrival rate λ/n and constant service times, $1/R$. Hence determine the maximum acceptable demand rate, λ, and the average response time. How do these results compare with the previous ones?

3. Consider a realisation (a sample path) of a queuing process where jobs arrive and depart singly. Let $N(t)$ be the number of jobs present at time t. This is a step function which jumps up by 1 at arrival instants and down by 1 at departure instants. Assume that $N(t)$ keeps reaching the value 0 from time to time.

Show that, with the possible exception of an initial period, to each step that $N(t)$ makes from j to $j + 1$, there corresponds a step from $j + 1$ to j, and vice versa ($j = 0, 1, \ldots$). Hence argue that, in the long run, the fraction of arrivals that see state j is equal to the fraction of departures that leave state j behind.

4. The server in an $M/G/1$ system in the steady-state goes through alternating periods of being busy (serving jobs) and being idle (no jobs to serve). Denote the average lengths of these periods by B and I, respectively. Clearly, the fraction of time that the server is busy is equal to $B/(B + I)$.

Show that, when the arrival process is Poisson, the average idle period is equal to the average interarrival time: $I = 1/\lambda$. From this, and from the known value of the server utilisation, derive the value of B.

5. Let us call a moment of time when there are j jobs in the system and one of them is just starting its service, a j-instant ($j = 1, 2, \ldots$). A 0-instant is a moment of departure after which the system is empty.

Show that the interval of time between a j-instant and the nearest subsequent $(j - 1)$ instant is statistically identical to a busy period. Hence determine the average length of the interval between a j-instant and the nearest subsequent 0-instant.

4.3 Priority scheduling

In many queuing systems of practical interest, the demand consists of jobs of different types. These job types may or may not have different arrival and service characteristics. Rather than treat them all equally and serve them in order of arrival, it is often desirable to discriminate among the different job types, giving a better quality of service to some, at the expense of others. The usual mechanism for doing that is to operate some sort of priority scheduling strategy. Let us examine a model of such a system.

Assume that there are K job types, numbered $1, 2, \ldots, K$. Type i jobs arrive in an independent Poisson stream with rate λ_i; their service requirements may have an arbitrary distribution function, $F_i(x)$, with mean $1/\mu_i$ and second moment M_{2i} ($i = 1, 2, \ldots, K$). There is a separate queue for each type, where jobs wait in order of arrival. Service is provided by a single server (figure 4.3).

Figure 4.3

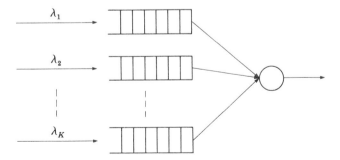

The different queues are attended by the server according to a priority assignment which we shall assume, for convenience, to be in inverse order of type indices. Thus, type 1 has the highest priority, type 2 the second highest, ..., type K the lowest. After a service completion, when a scheduling decision has to be made as to which job to start serving next, the job selected is the one at the head of the highest priority (lowest index) non-empty queue. This means, of course, that a type i job may start service only if queues $1, 2, \ldots, i - 1$ are empty ($i = 2, 3, \ldots, K$).

In order to complete the definition of the scheduling strategy, it remains to specify what happens if a higher priority job arrives and finds a lower priority one in service. One possibility is to take no action other than place the new arrival in its queue and await the scheduling decision that will be made on completing the current service. A strategy that does this is called a 'non-preemptive', or 'head-of-the-line' priority scheduling strategy. Alternatively, the new arrival may be allowed to interrupt the current service and occupy the server immediately. The displaced job goes back to the head of its queue. Such a strategy is said to be 'preemptive'. Preemptive strategies are further distinguished by the way they deal with an interrupted job. If, when that job eventually regains the server, its service is continued from the point of interruption, the strategy is called 'preemptive-resume'. If exactly the same service is restarted from the beginning, then the strategy is 'preemptive-repeat without resampling'. If the job starts a new service (i.e. a new instance of the same random variable) from the beginning, then the strategy is 'preemptive-repeat with resampling'.

We shall examine the non-preemptive and the preemptive-resume scheduling strategies. The object will be to determine the steady-state average queuing and response times for the different job types. This can be done by expressing these quantities in terms of themselves, in much the same way as for the $M/G/1$ queue. Other performance measures involving higher moments or distributions, and the other two preemptive scheduling strategies, would require a deeper analysis which is outside the scope of this book.

The following notation will be used (all probabilities and means refer to the steady-state):

$\rho_i = \lambda_i / \mu_i$: load for type i;
U_i: probability that a type i job is in service;
L_i: average number of type i jobs in the system;
W_i: average time in the system for a type i job (response time);
L_{Qi}: average number of type i jobs waiting in their queue;
W_{Qi}: average time in a queue for a type i job (waiting time).

The condition for non-saturation is that the total load should be less than 1: $\rho_1 + \rho_2 + \cdots + \rho_K < 1$. We shall assume that this is the case, and that the system is in the steady-state. For both the non-preemptive and the preemptive-resume scheduling strategies, the average time that a type i job spends in service is $1/\mu_i$. Therefore, while the server is serving type i jobs, the latter depart at the rate of μ_i per unit time. Hence, the departure rate for type i jobs is $U_i \mu_i$. Since this must be equal to the arrival rate, λ_i, we conclude that $U_i = \rho_i$ ($i = 1, 2, \ldots, K$).

Applying Little's result to the type i jobs in the system and in the queue, respectively, we get

$$L_i = \lambda_i W_i; \quad L_{Qi} = \lambda_i W_{Qi}; \quad i = 1, 2, \ldots, K. \tag{30}$$

Also, under these two scheduling strategies, the response time is equal to the waiting time plus the service time:

$$W_i = W_{Qi} + \frac{1}{\mu_i}; \quad i = 1, 2, \ldots, K. \tag{31}$$

Let us concentrate now on the non-preemptive scheduling strategy. The waiting time of a type i job can be represented as a sum of three distinct components. The first of these is waiting for the currently served job, if any, to complete. Denote the average of that wait by W_0. With probability ρ_j, a job of type j is in service. If so, its average remaining service time is $\mu_j M_{2j}/2$, according to 2.(11). Hence,

$$W_0 = \sum_{j=1}^{K} \frac{\rho_j \mu_j M_{2j}}{2} = \frac{1}{2} \sum_{j=1}^{K} \lambda_j M_{2j}. \tag{32}$$

The second component of the type i job waiting time is the delay imposed by all the jobs of equal or higher priority (i.e. types $1, 2, \ldots, i$), that are found in their queues on arrival. Denote the average of that second component by A_i. Since the average number of jobs in queue j is L_{Qj}, and each of them takes on the average $1/\mu_j$ to serve, we can write

$$A_i = \sum_{j=1}^{i} \frac{1}{\mu_j} L_{Qj} = \sum_{j=1}^{i} \rho_j W_{Qj}; \quad i = 1, 2, \ldots, K \tag{33}$$

(the second equality in (33) is a consequence of (30)).

Finally, there is the delay resulting from the higher priority jobs that arrive while the type i job is waiting (those jobs precede it at the server). During the average waiting time W_{Qi}, an average of $\lambda_j W_{Qi}$ jobs of type j arrive ($j = 1, 2, \ldots, i-1$). Each of them takes an average of $1/\mu_j$ to serve.

Therefore, the total additional delay, B_i, is given by

$$B_i = W_{Qi} \sum_{j=1}^{i-1} \rho_j; \quad i = 2, 3, \ldots, K. \tag{34}$$

Of course, the component B_i does not exist for $i = 1$.

Putting together (32), (33) and (34), we obtain a set of equations for the unknown average queuing times W_{Qi}:

$$W_{Qi} = W_0 + \sum_{j=1}^{i} \rho_j W_{Qj} + W_{Qi} \sum_{j=1}^{i-1} \rho_j; \quad i = 1, 2, \ldots, K \tag{35}$$

(an empty sum is equal to 0 by definition). This set of equations is triangular: the first equation contains W_{Q1} only, the second W_{Q1} and W_{Q2}, etc. The solution is easily found to be

$$W_{Qi} = \frac{W_0}{(1 - \sum_{j=1}^{i-1} \rho_j)(1 - \sum_{j=1}^{i} \rho_j)}; \quad i = 1, 2, \ldots, K, \tag{36}$$

where W_0 is given by (32). Expressions (36) are known as 'Cobham's formulae'. The average response times, W_i, and numbers of jobs, L_i and L_{Qi}, are obtained from (30) and (31).

Certain implications of these results are worth pointing out. First, the average waiting (and response) time for type i is influenced by the lower priority types $(i + 1, i + 2, \ldots, K)$ only through the expected residual service delay, W_0. It is obvious from the derivation of that quantity that it is in fact independent of the scheduling strategy, provided that the latter does not allow service interruptions. The higher priority job types $(1, 2, \ldots, i - 1)$ influence type i through their total load $(\rho_1 + \rho_2 + \cdots + \rho_{i-1})$, as well as through W_0. That total load is also independent of the scheduling strategy. Therefore, as far as the type i jobs are concerned, all jobs can be separated into three essential groups: group 1, which comprises types $1, 2, \ldots, i - 1$; group 2, which has type i on its own; and group 3, which consists of types $i + 1, i + 2, \ldots, K$. As long as group 2 has higher priority than group 3, and lower priority than group 1, the order of service within groups 1 and 3 is immaterial. Any non-preemptive scheduling strategy can be employed there (e.g. FIFO), without affecting type i.

A look at the denominator in (36) convinces us that, in order for W_{Qi} to be finite, it is sufficient that the total load for types $1, 2, \ldots, i$ is less than 1. Suppose, for instance, that $\rho_1 + \rho_2 + \cdots + \rho_i < 1$, but $\rho_1 + \rho_2 + \cdots + \rho_i + \rho_{i+1} > 1$. Then the first i queues are non-saturated, while the remaining $K - i$ ones are saturated. The probability that a job of type $i + 1$ is in service is equal to $1 - \rho_1 - \rho_2 - \cdots - \rho_i$. Jobs of types

$i + 2, \ldots, K$ are never in service, in the long run. Cobham's formulae continue to hold up to, and including, type i. However, the expression (32) for W_0 should be replaced by

$$W_0 = \frac{1}{2}\left[\sum_{j=1}^{i} \lambda_j M_{2j} + (1 - \rho_1 - \rho_2 - \cdots - \rho_i)\mu_{i+1}M_{2,i+1} \right].$$

The average waiting times for the saturated job types are infinite.

Example

Jobs arrive into a single-server system according to a Poisson stream with rate λ. There are K possible job lengths: the required service time of a job is equal to x_i with probability f_i ($i = 1, 2, \ldots, K$; $f_1 + f_2 + \cdots + f_K = 1$), where $x_1 < x_2 < \cdots < x_K$. These required service times are known on arrival and the scheduling strategy gives non-preemptive priority to shorter jobs over longer ones. Thus, the jobs that require time x_1 have top priority, those that require x_2 are second, $\ldots$, the ones that require x_K are bottom. This scheduling strategy is called 'shortest-processing-time-first', or SPT.

When is the system non-saturated and what is the average waiting time for the jobs that require service time x_i ($i = 1, 2, \ldots, K$)?

Here we have a priority model where the type of a job is determined by its length, or required service time. From the decomposition property of the Poisson process (section 2.2) it follows that the jobs of type x_i arrive in a Poisson stream with rate $f_i\lambda$. The first and second moments of their service times are, of course, x_i and x_i^2, respectively. The load corresponding to the ith type is $\lambda f_i x_i$ ($i = 1, 2, \ldots, K$). The condition for non-saturation of the entire system is

$$\lambda \sum_{i=1}^{K} f_i x_i < 1.$$

When the system is in the steady-state, the average residual service delay is equal to

$$W_0 = \frac{\lambda}{2} \sum_{i=1}^{K} f_i x_i^2 = \frac{\lambda}{2} M_2,$$

where M_2 is the second moment of the required service time.

The average waiting time for jobs of type x_i, $W_Q(x_i)$, is obtained from Cobham's formulae:

$$W_Q(x_i) = \frac{W_0}{(1 - \lambda \sum_{j=1}^{i-1} f_j x_j)(1 - \lambda \sum_{j=1}^{i} f_j x_j)}; \quad i = 1, 2, \ldots, K.$$

It is also possible for some job types to be saturated, while others are not. In that case, the expression for W_0 has to be modified appropriately.

Let us consider now the steady-state performance of the preemptive-resume priority scheduling strategy. The service of a lower priority job may be interrupted by the subsequent arrival of a higher priority one. The preempted job is returned to the head of its queue and eventually resumes its service from the point of interruption. This may happen many times before the service is completed.

Define the 'initial waiting time' of a job as the period between its arrival and the start of its service. That is followed by what we shall call the 'attendance time', which is the period between the start and the completion of the service. The attendance time may contain further waits caused by preemptions. Denote the average initial waiting time and the average attendance time of a type i job by T_i and V_i, respectively.

The following two observations will help us to determine T_i. First, the jobs of types $i + 1, i + 2, \ldots, K$ may be ignored when considering the performance of a type i job. This is because the latter can preempt the former and so cannot be delayed by them. Second, during the initial waiting time of a type i job, it does not matter whether the priorities of types $1, 2, \ldots, i - 1$ are preemptive or non-preemptive. In both cases, the job starts service when there are no type i jobs ahead of it and no higher priority jobs in the system.

We conclude, therefore, that the average initial waiting time of a type i job under a preemptive priority strategy is equal to the average waiting time of a type i job under a non-preemptive priority strategy, in a system where types $i + 1, i + 2, \ldots, K$ do not exist. The corresponding Cobham formula provides the desired expression for T_i:

$$T_i = \frac{W_{0i}}{(1 - \sum_{j=1}^{i-1} \rho_j)(1 - \sum_{j=1}^{i} \rho_j)}; \quad i = 1, 2, \ldots, K, \tag{37}$$

where W_{0i} is the residual service delay involving types $1, 2, \ldots, i$ only:

$$W_{0i} = \frac{1}{2} \sum_{j=1}^{i} \lambda_j M_{2j}; \quad i = 1, 2, \ldots, K. \tag{38}$$

Next, consider the attendance time of a type i job, V_i. This consists of the job's own service time, plus the service times of all higher priority jobs that arrive during the attendance time. During the period V_i, an average of $\lambda_j V_i$ jobs of type j arrive; each of them takes an average of $1/\mu_j$ to serve. Hence, we can write

$$V_i = \frac{1}{\mu_i} + \sum_{j=1}^{i-1} \frac{\lambda_j}{\mu_j} V_i; \quad i = 1, 2, \ldots, K. \tag{39}$$

Solving this for V_i yields

$$V_i = \frac{1}{\mu_i(1 - \sum_{j=1}^{i-1} \rho_j)}; \quad i = 1, 2, \ldots, K. \tag{40}$$

The average response time, W_i, and total waiting time, W_{Qi}, for type i are given by

$$W_i = T_i + V_i; \quad W_{Qi} = T_i + V_i - \frac{1}{\mu_i}; \qquad i = 1, 2, \ldots, K. \tag{41}$$

Average numbers of jobs of different types in the system and in the queues are obtained from (30).

The overall average performance of either the preemptive or the non-preemptive scheduling strategy may also be of interest. This is easily derived from the available results. The total average number of jobs in the system, L, is a straightforward sum

$$L = \sum_{i=1}^{K} L_i. \tag{42}$$

Let $\lambda = \lambda_1 + \lambda_2 + \cdots + \lambda_K$ be the total arrival rate. Little's result, applied to the whole system, yields the overall average response time, W:

$$W = \frac{L}{\lambda} = \frac{1}{\lambda} \sum_{i=1}^{K} L_i = \frac{1}{\lambda} \sum_{i=1}^{K} \lambda_i W_i. \tag{43}$$

The last equation has a simple probabilistic interpretation. Since an average of λ_i jobs of type i arrive into the system per unit time, λ_i/λ is the fraction of all arrivals that are of type i. Alternatively, it is the probability that a new arrival is of type i. The right-hand side of (43) is thus the appropriate weighted average of response times.

Exercises

1. The jobs in a single-processor computer system are of two types. They arrive in Poisson streams, with rates 5 and 10 jobs per minute, respectively. The required service times for both types are exponentially distributed, but with different means: 6 seconds and 2 seconds, respectively.

Find the average response times for the two job types, and the overall average response time, under the following scheduling strategies:

(a) type 1 has non-preemptive priority over type 2;

(b) type 1 has preemptive-resume priority over type 2;

(c) type 2 has non-preemptive priority over type 1;

(d) type 2 has preemptive-resume priority over type 1;

(e) all jobs join a single queue and are served in FIFO order, regardless of type.

(When the two job types are merged together, the resulting service time density function is

$$f(x) = \frac{\lambda_1}{\lambda} \mu_1 e^{-\mu_1 x} + \frac{\lambda_2}{\lambda} \mu_2 e^{-\mu_2 x},$$

where $\lambda = \lambda_1 + \lambda_2$. Under the FIFO scheduling strategy, the average waiting time, W_Q, is the same for all jobs and is given by the appropriate $M/G/1$ result. The average response times for types 1 and 2 are equal to $W_Q + (1/\mu_1)$ and $W_Q + (1/\mu_2)$, respectively.)

2. Generalise the model from the example in this section by assuming that the service times have some general distribution function, $F(x)$. The scheduling strategy is shortest-processing-time-first (non-preemptive). That is, a job requiring service time x is selected for service before a job requiring time y if $x < y$. Show that the expected residual service delay, W_0, is again equal to $(\lambda/2) M_2$, where M_2 is the second moment of the service time. Using the required service time of a job as a type identifier and replacing the sums in Cobham's formulae by integrals, derive the following expression of the average waiting time, $W_Q(x)$, of a job whose required service time is x:

$$W_Q(x) = \frac{W_0}{[1 - \lambda \int_0^{x-} u\,dF(u)][1 - \lambda \int_0^{x} u\,dF(u)]}; \quad x > 0,$$

where $x-$ indicates limit from the left. If the distribution function F is continuous at point x, then the two terms in square brackets are equal.

4.4   The processor-sharing strategy

In computer systems where a single processor provides service to a number of jobs running concurrently, the following scheduling strategy is usually employed. The processor offers service in quanta of fixed size, Q. The job at the head of the queue is given a quantum of service. If that job completes before the quantum elapses, it departs immediately; otherwise it returns to the end of the queue and awaits its turn again. Thus, if there are n jobs competing for the processor, each of them occupies it for one out of every n quanta. This scheduling strategy is called 'round-robin'; it is illustrated in figure 4.4.

Figure 4.4

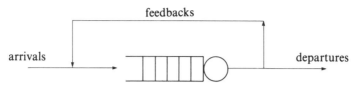

The round-robin scheduling strategy can be analysed, under suitable assumptions, by following the progress of a job as it circulates round the system. Such an analysis tends to be rather involved, because the successive passes that the job makes through the queue are not independent of each other. Much of the complexity disappears, however, if the quantum size Q is allowed to shrink to 0. The smaller the quantum, the higher the frequency with which each of the waiting jobs visits the processor. In the limit $Q \to 0$, the picture of frantically circulating jobs blurs into one where all competing jobs run smoothly in parallel, each receiving an equal fraction of the processing capacity. That scheduling strategy is called 'processor-sharing'. It is a good approximation of the round-robin with a small quantum, and is considerably easier to analyse.

The processor-sharing strategy can be defined directly by saying that if there are n jobs in the system at time t then in the infinitesimal interval $(t, t + dt)$, each of those n jobs receives an amount of service equal to dt/n. Alternatively, the time necessary for every one of the n jobs to increase its attained service by an infinitesimal amount, ds, is equal to nds.

We shall assume that jobs arrive in a Poisson stream with rate λ, and that their required service times have some general distribution function, $F(x)$, with mean $1/\mu$. The object of interest is the steady-state average response time for a job whose required service time is x. That average will be denoted by $W(x)$. The unconditional steady-state average response time, W, is obtained in terms of $W(x)$ according to

$$W = \int_0^\infty W(x) \, dF(x). \tag{44}$$

The steady-state average number of jobs in the system, L, is related to W by Little's result: $L = \lambda W$.

Steady-state exists if the load, $\rho = \lambda/\mu$, is less than 1. This is assumed to be the case.

The above performance measures can be determined in a very straightforward manner, by arguing as follows. Let J be a job whose service requirement is x. On arrival, J finds an average of L jobs already present (random observer property of the Poisson process). Since the system is in steady-state, J shares the processor with L other jobs, on the average, until it is completed. On the other hand, the time that it takes for a job to attain service x when $L + 1$ jobs share the processor, is $(L + 1)x$. Hence,

$$W(x) = (L + 1)x = (\lambda W + 1)x. \tag{45}$$

Substituting (45) into (44), we get an equation for the unconditional average response time, W:

$$W = (\lambda W + 1) \int_0^\infty x \, dF(x) = (\lambda W + 1) \frac{1}{\mu}. \tag{46}$$

This yields

$$W = \frac{1}{\mu(1 - \rho)}. \tag{47}$$

Going back to (45), the expression for $W(x)$ becomes

$$W(x) = \frac{x}{(1 - \rho)}. \tag{48}$$

The weak point in this argument is the assertion (not quite an obvious one) that the average number of jobs with which J shares the processor remains constant throughout J's residence in the system. However, both that assertion and the resulting formulae are correct and can be established rigorously.

We have reached the rather remarkable conclusion that both the unconditional and the conditional average response times depend only on the mean required service time and not on the shape of the distribution, $F(x)$. This insensitivity extends to the distribution of the number of jobs in the system. It turns out that the latter is the same as for the M/M/1 queue (we state this without proof):

$$p_j = P(j \text{ jobs in the system}) = \rho^j(1 - \rho); \qquad j = 0, 1, \dots . \tag{49}$$

The most obvious implication of the result (48) is that the processor-sharing strategy favours short jobs at the expense of long ones. That is precisely the effect one normally wishes to achieve in an interactive computing environment. Consider, for example, the M/M/1 queue with FIFO scheduling. The average response time for a job requiring service x in that system is given by

$$W_{\mathrm{M/M/1}}(x) = \frac{\rho}{\mu(1 - \rho)} + x.$$

(The first term in the right-hand side is the average waiting time.) Comparing this with (48), we see that for the jobs whose required service time satisfies $x < 1/\mu$ (i.e. the ones that are shorter than average), the processor-sharing strategy is preferable to FIFO. Conversely, if $x > 1/\mu$, then FIFO is better.

When the required service times are distributed exponentially, processor-sharing and FIFO have the same unconditional average response time, W. In the general case, we can compare (47) with the corresponding M/G/1 result,

$$W_{\mathrm{M/G/1}} = \frac{1}{\mu}\left[1 + \frac{\rho(1 + C^2)}{2(1 - \rho)}\right].$$

The outcome of the comparison depends on the squared coefficient of variation, C^2. If that is greater than 1, then processor-sharing is better; otherwise, FIFO is better (for the exponential distribution, $C^2 = 1$).

Suppose now that the job population consists of K job types, arriving in independent Poisson streams with possibly different rates (λ_i for type i), and having different distributions of required service times ($F_i(x)$ for type i; mean $1/\mu_i$). The scheduling strategy is again processor-sharing.

Because of the insensitivity of the average response times (conditional and unconditional) with respect to the service time distribution, expressions (48) and (47) continue to hold. The overall average service time, $1/\mu$, and the total load, ρ, are given by

$$\frac{1}{\mu} = \sum_{i=1}^{K} \frac{\lambda_i}{\lambda} \frac{1}{\mu_i} = \frac{1}{\lambda}\sum_{i=1}^{K} \rho_i; \quad \rho = \sum_{i=1}^{K} \rho_i, \tag{50}$$

where $\lambda = \lambda_1 + \lambda_2 + \cdots + \lambda_K$ is the total arrival rate (λ_i/λ is the probability that an incoming job is of type i). The average response time for a job of type i, W_i, is obtained from

$$W_i = \int_0^\infty W(x)\mathrm{d}F_i(x) = \frac{1}{\mu_i(1 - \rho)}; \quad i = 1, 2, \ldots, K. \tag{51}$$

Note that the influence of the other job types is expressed only through their total load. The average number of type i jobs in the system, L_i, is given by Little's result

$$L_i = \lambda_i W_i = \frac{\rho_i}{1 - \rho}; \quad i = 1, 2, \ldots, K. \tag{52}$$

The steady-state joint distribution of the numbers of jobs of different types is also easy to determine. Denote by $p(j_1, j_2, \ldots, j_K)$ the probability that there are j_1 jobs of type 1, j_2 jobs of type 2, $\ldots$, j_K jobs of type K in the system. Furthermore, let $p(j_1, j_2, \ldots, j_K \mid j)$ be the conditional probability of the same event, given that the total number of jobs in the system is j ($j = j_1 + j_2 + \cdots + j_K$). Remembering that the distribution of the total number is given by (49), we can write

$$p(j_1, j_2, \ldots, j_K) = \rho^j (1 - \rho) p(j_1, j_2, \ldots, j_K \mid j). \tag{53}$$

Now, it follows from (52) that the fraction of type i jobs in the system, L_i/L, is equal to ρ_i/ρ. That ratio can also be interpreted as the probability that a job observed in the system is of type i ($i = 1, 2, \ldots, K$). This, together with the fact that the types of the jobs in the system are independent of each other, implies that, if there are j jobs present, the probability that a given j_1 of them are of type 1, a given j_2 of them are of type 2, ..., a given j_K of them are of type K, is equal to

$$\left(\frac{\rho_1}{\rho}\right)^{j_1} \left(\frac{\rho_2}{\rho}\right)^{j_2} \cdots \left(\frac{\rho_K}{\rho}\right)^{j_K} = \frac{1}{\rho^j}\, \rho_1^{j_1} \rho_2^{j_2} \cdots \rho_K^{j_K}.$$

Hence, the conditional probability which appears in (53) is given by

$$p(j_1, j_2, \ldots, j_K \mid j) = \frac{j!}{j_1! \, j_2! \cdots j_K!} \left(\frac{1}{\rho^j}\right) \rho_1^{j_1} \rho_2^{j_2} \cdots \rho_K^{j_K} \tag{54}$$

(the coefficient in the right-hand side represents the number of ways of splitting a set of j jobs into K groups of sizes $j_1, j_2, \ldots, j_K$ respectively).

Substituting (54) into (53) we get the desired joint probability:

$$p(j_1, j_2, \ldots, j_K) = (1 - \rho) j! \left(\frac{\rho_1^{j_1}}{j_1!}\right) \left(\frac{\rho_2^{j_2}}{j_2!}\right) \cdots \left(\frac{\rho_K^{j_K}}{j_K!}\right). \tag{55}$$

From this result, a straightforward manipulation leads to the marginal distribution of the number of type i jobs in the system. The latter turns out to be geometric:

$$P(j \text{ jobs of type } i \text{ in the system}) = \sigma_i^j (1 - \sigma_i);$$

$$j = 0, 1, \ldots, \quad (56)$$

where $\sigma_i = \rho_i/(1 - \rho + \rho_i)$. We leave the details of the derivation to the reader. Note that, although the joint distribution (55) has the form of a product, it is not equal to the product of the marginal distributions. That is, the numbers of jobs of different types in the system are not independent of each other.

We have seen that many results associated with the processor-sharing strategy are distinguished by their simplicity and by their insensitivity to the shape of the required service time distributions. The reason for these, and other nice properties (some of which will be mentioned in the next chapter), has to do with the egalitarian way in which jobs are treated: at any moment in time, all jobs in the system receive equal fractions of the processing capacity, regardless of length, type, or any other attributes. If that principle is abandoned, for example by allocating a larger fraction of capacity to one job type and a smaller fraction to another, then none of the present results would apply. The analysis of a non-egalitarian processor-sharing strategy is considerably more complex than that of a system with priorities.

Exercises

1. Jobs arrive into a single processor system in a Poisson stream, at the rate of 8 per minute. The required service time of a job is either 3 seconds, with probability 0.99, or 5 minutes, with probability 0.01. Two scheduling strategies are being considered: processor-sharing and FIFO. Compare the performance of those strategies in terms of (i) the average response time for the 3-second jobs, (ii) the average response time for the 5-minute jobs and (iii) the overall average response time.

2. Consider a processor-sharing system with arrival rate λ and density function of the required service times $f(x)$. Let $h(x)$ be the density function of the required service time of a job in the system, in the steady-state. Find $h(x)$ (as a ratio of loads), and show that it is the same as the density function of a randomly observed renewal period, in a renewal process where the renewal period density is $f(x)$.

3. From the definition of the conditional average response time it follows that $dW(x) = W(x + dx) - W(x)$ is the average time needed for a job whose required service time is greater than x, to increase its attained service from x to $x + dx$. Let $n(x)$ be the average number of jobs in the system whose attained service is less than or equal to x. Then $dn(x) = n(x + dx) - n(x)$ is the average number of jobs in the system whose attained service is in the interval $(x, x + dx)$. By applying Little's result to those jobs, show that

$$dn(x) = \lambda[1 - F(x)]dW(x),$$

where $F(x)$ is the distribution function of the required service times. Hence derive an expression for $n(x)$ in the case of the processor-sharing strategy.

4.5   Optimal scheduling strategies
The factors that influence the performance of a computer system can be grouped into three broad categories. These are (i) the hardware characteristics (e.g. speed of processors and I/O devices), (ii) the demand characteristics (e.g. type and arrival pattern of jobs) and (iii) the scheduling strategy or strategies employed. The first two categories are usually outside the day-to-day control of the system management, but the scheduling

strategy may be easy to change. One is then faced with the question, which is the best strategy to use?

We shall pose that question in the context of the family of non-preemptive priority strategies. The assumptions are as in section 4.3. The demand consists of K job types, arriving in independent Poisson streams and having general service requirements. The parameters for type i are λ_i (arrival rate), $1/\mu_i$ (average service time) and M_{2i} (second moment of the service time). There is a single server, and the scheduling strategy is constrained to be of the non-preemptive priority kind.

We now have a choice of $K!$ possible strategies: each permutation of the indices $\{1, 2, \ldots, K\}$ can be used to specify which job type should have top priority, which second top, etc. In section 4.3, the priority assignment was defined by the permutation $(1, 2, \ldots, K)$.

The object is to minimise a cost function of the form

$$C = \sum_{i=1}^{K} c_i W_{Qi}, \tag{57}$$

where W_{Qi} is the average waiting time for jobs of type i, and c_i is a non-negative constant. The latter reflects the relative importance attached to type i. We could have chosen as a cost function a linear combination of average response times, rather than average waiting times. However, since $W_i = W_{Qi} + (1/\mu_i)$, the two objectives are equivalent.

Intuitively, if c_i is large, then in order to keep the cost low, W_{Qi} should be small, i.e. type i should be given high priority. However, a priority ordering based solely on the magnitudes of the coefficients c_i is not necessarily optimal. The loads $\rho_i = \lambda_i/\mu_i$ also play a role in determining the best strategy, since they influence the waiting times.

In principle, one could use Cobham's formulae (36) to evaluate the cost function for each of the $K!$ priority assignments, and then compare the results. In practice, of course, such an approach cannot be contemplated, except for a few very small values of K. Fortunately, the optimal scheduling strategy can be found directly, by examining the effect of local changes in the order of priorities.

First, we shall establish a relation between the K average waiting times which, besides being the key to our argument, is of interest in its own right. That relation has the following form:

$$\sum_{i=1}^{K} \rho_i W_{Qi} = \frac{\rho W_0}{1 - \rho}, \tag{58}$$

where $\rho = \rho_1 + \rho_2 + \cdots + \rho_K$ is the total load and W_0 is the expected residual service delay, given by (32). The proof of (58) in the case that we are considering is immediate: it suffices to write expression (35) for $i = K$, and substitute in it the corresponding Cobham formula (36) for W_{QK}. In fact, (58) has a much more general validity. It holds, under mild restrictions, for any scheduling strategy which does not allow interruptions of services.

Note that the right-hand side of (58) is completely independent of the assignment of priorities to job types. Therefore, any changes in the average response times, resulting from a change of scheduling strategy, must be such that the value of the linear combination in the left-hand side of (58) remains constant. We say that the above linear combination is conserved within the family of non-preemptive priority scheduling strategies (also, within a much larger family of scheduling strategies without service interruptions).

The result (58) is usually referred to as 'Kleinrock's conservation law'.

Let us examine now the way in which the cost function varies with changes in the priority assignment. Two scheduling strategies are said to be 'neighbouring', if they differ only in the priority order of two adjacent job types. For instance, in a system with five job types, the strategies $(2, 4, 5, 3, 1)$ and $(2, 5, 4, 3, 1)$ are neighbouring. Let A and B be two neighbouring scheduling strategies, and i and j be the adjacent job types that swap priorities: $A = (\ldots, i, j, \ldots)$; $B = (\ldots, j, i, \ldots)$. Denote the average waiting time of a type s job under A and B by W_{Qs}^A and W_{Qs}^B respectively $(s = 1, 2, \ldots, K)$. The corresponding values of the cost function are C^A and C^B, respectively.

From the properties of the priority scheduling strategies it follows that all job types except i and j have the same average waiting times under A and B (see the remark following (36)). Hence, the difference in the values of the cost function is equal to

$$
\begin{aligned}
C^A - C^B &= (c_i W_{Qi}^A + c_j W_{Qj}^A) - (c_i W_{Qi}^B + c_j W_{Qj}^B) \\
&= c_i(W_{Qi}^A - W_{Qi}^B) + c_j(W_{Qj}^A - W_{Qj}^B)
\end{aligned}
\tag{59}
$$

(all the other terms cancel out). Also, since the linear combination in the left-hand side of (58) has the same value under A and B, we must have

$$
\rho_i W_{Qi}^A + \rho_j W_{Qj}^A = \rho_i W_{Qi}^B + \rho_j W_{Qj}^B
\tag{60}
$$

(again, all the other terms cancel out). This last equality can be rewritten as

$$
W_{Qi}^A - W_{Qi}^B = -\frac{\rho_j}{\rho_i}(W_{Qj}^A - W_{Qj}^B).
\tag{61}
$$

After substitution of (61), expression (59) becomes

$$
C^A - C^B = (W_{Qj}^A - W_{Qj}^B)\left(c_j - c_i\frac{\rho_j}{\rho_i}\right).
\tag{62}
$$

Note that the first factor in the right-hand side of (62) is always positive. This is because type j has higher priority, and hence lower average waiting time, under strategy B than under A. Therefore, the sign of the whole expression is determined by that of the second factor. This implies that strategy B is better than strategy A if, and only if,

$$\frac{c_i}{\rho_i} < \frac{c_j}{\rho_j}. \tag{63}$$

It is now easy to construct a scheduling strategy that minimises the cost function. We shall do this by first re-numbering the job types in such a way that the following inequalities are satisfied:

$$\frac{c_1}{\rho_1} \geqslant \frac{c_2}{\rho_2} \geqslant \cdots \geqslant \frac{c_K}{\rho_K}. \tag{64}$$

Then the optimal strategy is to give top priority to type 1, second top to type 2, ..., bottom priority to type K.

Indeed, if the strategy $(1, 2, \ldots, K)$ was not optimal, then one of its neighbouring strategies, or their neighbouring strategies, etc., would be better than it. That, however, is impossible, in view of (64).

Example

Suppose that the coefficients of the cost function are (or are proportional to) the arrival rates of the corresponding job types:

$$C = \sum_{i=1}^{K} \lambda_i W_{Qi}. \tag{65}$$

Remembering that an incoming job is of type i with probability λ_i/λ, where λ is the total arrival rate, we see that minimising (65) is equivalent to minimising the overall average waiting time, W_Q (see equation (43)).

Since $\lambda_i/\rho_i = \mu_i$, the ordering (64) is determined solely by the service time parameters. An optimal priority assignment is one which satisfies

$$\mu_1 \geqslant \mu_2 \geqslant \cdots \geqslant \mu_K,$$

or, in terms of the average required service times,

$$\frac{1}{\mu_1} \leqslant \frac{1}{\mu_2} \leqslant \cdots \leqslant \frac{1}{\mu_K}. \tag{66}$$

In other words, jobs with lower average service times should be given higher priority. This scheduling strategy is referred to as 'shortest-expected-processing-time-first', or SEPT. As long as preemptions are disallowed, and the only information available about a job is its type (i.e.

the distribution of its required service time), the strategy that minimises the overall average waiting time is SEPT. The case where the exact required service times of all jobs are known on arrival can be treated as a special case of this result (see exercise 1).

Thus, in the non-preemptive case, optimal scheduling decisions can be taken without worrying about the distributions of the required service times. Only the means are important. However, if preemptive strategies are also taken into consideration, the situation becomes considerably more complicated.

Suppose that, at time 0, there are K jobs at the server, and that the required service time of job i has distribution function $F_i(x)$ $(i = 1, 2, \ldots, K)$. There are no further arrivals and services can be interrupted at arbitrary points. The object is to find a scheduling strategy that minimises a cost function of the form

$$C = \sum_{i=1}^{K} c_i W_i, \tag{67}$$

where W_i is the average completion time of job i.

The solution to this problem was found fairly recently. It is not an intuitively obvious one and is quite difficult to prove. We shall present the relevant results without attempting to justify them.

Suppose that job i has already attained service x and that the server is assigned to it for a period y. The conditional probability that the job will complete within that period is

$$S_i(x, y) = \frac{F_i(x + y) - F_i(x)}{1 - F_i(x)}; \quad x \geqslant 0, y > 0. \tag{68}$$

The average amount of service that the job will actually use during the period y is equal to

$$Q_i(x, y) = \frac{\int_0^y [1 - F_i(x + u)]\,du}{1 - F_i(x)}; \quad x \geqslant 0, y > 0. \tag{69}$$

If the completion of a job is regarded as a desirable occurrence which is rewarded, then the ratio $S_i(x, y)/Q_i(x, y)$ can be thought of as the expected reward from job i per unit of processor time used during the period y. In general, that ratio varies with y. Let $v_i(x)$ be the largest value that it can reach, over all possible y:

$$v_i(x) = \max_{y > 0} \frac{S_i(x, y)}{Q_i(x, y)}; \quad x \geqslant 0. \tag{70}$$

The quantity $v_i(x)$ is called the 'index', or 'rank', of job i when the latter has attained service x. The smallest value of y for which the maximum in (70) is reached is called the 'rank quantum'. The optimal scheduling strategy can now be described as follows:

At each scheduling decision instant (the first being at time 0), select the job for which the product $c_i v_i(x)$ is largest. Assign the server to that job for the duration of the corresponding rank quantum. The end of the quantum, or the completion of the job if the latter occurs earlier, defines the next decision instant. Clearly, at most one rank has to be re-computed then: the one of the job that had the server (if it has not completed). The attained service times of all other jobs remain unchanged.

We shall refer to this as the 'dynamic rank', or DR scheduling strategy. It can be shown that if, instead of being all present at the beginning, jobs of different types arrive in independent Poisson streams, the optimal scheduling strategy is still, essentially, DR. Consider again the model with K job types characterised by their arrival rates, λ_i, and service times distribution functions, $F_i(x)$ $(i = 1, 2, \ldots, K)$. The cost function to be minimised is either (57) or a similar linear combination of average response times. Arbitrary preemptions are allowed.

The DR scheduling strategy is optimal in this case, provided that it is modified in the following two particulars.

(i) Scheduling decisions are made at job arrival instants, as well as at completions and rank quanta terminations.

(ii) The job selected for service is the one for which the quantity $(c_i/\lambda_i)v_i(x)$ is largest, where i is the job type, x is its attained service time and $v_i(x)$ is the rank computed according to (70).

Note the analogy between the non-preemptive scheduling rule, (64), and the DR one, (ii). The role of the average service time, $1/\mu_i$, is now played by the reciprocal of the rank, $1/v_i(x)$. Of course, the latter may change during the residence of a job in the system, as it attains portions of its required service.

Two important special cases of the DR strategy are exemplified by the quantities

$$r_i(x) = \frac{f_i(x)}{1 - F_i(x)}; \quad x \geqslant 0, \ i = 1, 2, \ldots, K. \tag{71}$$

where $f_i(x)$ is the type i required service time density function. The right-hand side of (71) represents the completion rate of a type i job which has attained service x.

If $r_i(x)$ is a monotone increasing function of x, then any allocation of the server to a type i job will have the effect of increasing its rank. Hence, the maximum in (70) is reached for $y = \infty$. An infinite rank quantum implies that the service of a type i job can be interrupted only by a new arrival of a higher rank, and not by any job already in the system. If this is true for all job types, then DR behaves like a preemptive priority strategy with priorities depending on the expected remaining service times.

If $r_i(x)$ is a decreasing function of x, then giving more service to a type i job decreases its rank. The maximum in (70) is reached for $y = 0$. In that case, $v_i(x) = r_i(x)$. The effect of an infinitesimal rank quantum is to serve the highest ranking job until its rank drops to the level of the next highest one. As soon as there are at least two jobs with equal (highest) rank, the strategy turns into processor-sharing between them.

The completion rate tends to be an increasing (decreasing) function of x when the corresponding coefficient of variation of the required service time is small (large). This, together with the above observations, leads us to a rather general scheduling principle: if the required service times are reasonably predictable, then having decided to serve a job, it should not be interrupted too much. If, on the other hand, the service times are very variable, frequent interruptions are better. Another manifestation of the same principle was observed in the comparison between the FIFO and processor-sharing strategies in section 4.4.

Before leaving this topic, a few words are in order about a problem which is closely related to optimisation. This concerns the characterisation of achievable performance. Suppose, for instance, that in our single-server system with K job types, performance is measured by the vector of average response times for the different types, $(W_1, W_2, \ldots, W_K)$. What is the set of such vectors that can be achieved by varying the scheduling strategy? Or, to put it another way, if a given vector of response times is specified as a performance objective, is there a scheduling strategy that can achieve it?

The general characterisation problem is still open. There are, however, solutions in some special cases where a conservation law like (58) applies. Let us examine, as an illustration, the case of two job types and consider the performance vectors, (W_1, W_2), achievable by non-preemptive scheduling strategies. Those vectors can be represented as points in the first quadrant of the two-dimensional plane.

The conservation law which applies to non-preemptive strategies tells us that the achievable points lie on a straight line defined by

$$\rho_1 W_1 + \rho_2 W_2 = C, \tag{72}$$

where C is a known constant (it is in fact given by the right-hand side of (58), for $K = 2$, plus the linear combination $(\rho_1/\mu_1) + (\rho_2/\mu_2)$, which accounts for the average service times). There are two special points on that line, corresponding to the two non-preemptive priority strategies: one where type 1 has top priority and one where type 2 has top priority (see figure 4.5).

Figure 4.5

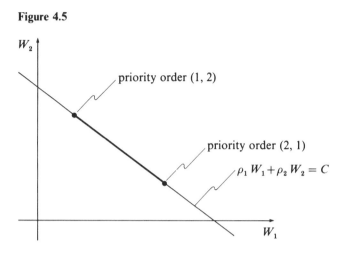

Those two points (whose coordinates are obtained with the aid of Cobham's formulae, section 4.3) are the extremes of the set of achievable performance vectors. The priority strategy $(1, 2)$ yields the lowest possible (without preemptions) average response time for type 1 and the highest possible one for type 2; the situation is reversed with the strategy $(2, 1)$. That is, no point to the left of $(1, 2)$, or to the right of $(2, 1)$, can be achievable. On the other hand, it can be demonstrated that all points between the two extremes are achievable. Thus, the set of achievable performance vectors coincides with the interval defined by $(1, 2)$ and $(2, 1)$.

In the case of K job types, there are $K!$ extreme points, corresponding to the $K!$ non-preemptive priority strategies. The set of achievable performance vectors is then a $(K - 1)$-dimensional polyhedron which has those points as its vertices (see figure 4.6 for an example when $K = 3$).

Figure 4.6

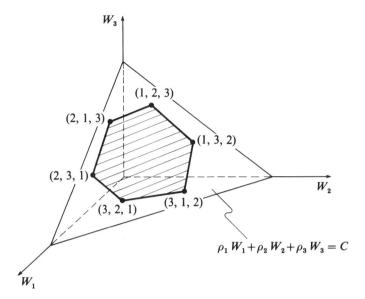

That structure implies that the maximum, or minimum, of any linear objective function of average response times is reached at one of the vertices. In other words, if a linear objective function is to be optimised by a non-preemptive scheduling strategy, the optimum is achieved by one of the $K!$ priority strategies. One important consequence of this result is that the non-preemptive priority strategy defined by the rule (64) is optimal, not only among the $K!$ members of its family, but among all non-preemptive scheduling strategies.

Exercises

1. Jobs arrive into a single-server system in a Poisson stream with rate λ; the distribution function of their service requirements is $F(x)$. The exact required service time of every job is known on arrival. The object is to minimise the overall average waiting (or response) time, using a non-preemptive scheduling strategy.

Treating the required service time as a type identifier and applying the result from the example in this section, show that the strategy which gives non-preemptive priority to shorter jobs over longer ones is optimal. That strategy is called 'shortest-processing-time first' (or SPT).

2. In a single-server system with K job types and Poisson arrivals, the required service times for type i are uniformly distributed on the interval $(0, b_i)$ $(i = 1, 2, \ldots, K)$. Show that the rank of a type i job with attained service time x is equal to

$$v_i(x) = \frac{2}{b_i - x}; \quad 0 < x < b_i, \, i = 1, 2, \ldots, K,$$

and that the corresponding rank quantum is $y = b_i - x$. Hence establish that a strategy that minimises the overall average response time is the preemptive 'shortest-expected-remaining-processing-time first' (or SERPT) strategy.

3. In a single-server system with K job types and Poisson arrivals, the required service times for type i have a 'hyperexponential' distribution:

$$F_i(x) = 1 - \alpha_i e^{-\mu_i x} - (1 - \alpha_i) e^{-v_i x};$$

$$0 < \alpha_i < 1, \, \mu_i, v_i > 0, \, i = 1, 2, \ldots, K.$$

Show that the completion rate $r_i(x)$ is a decreasing function of x, for all job types. Hence determine a scheduling strategy (which may use preemptions) that minimises a cost function of the form (57).

Literature

Little's result had existed as a 'folk theorem' for a long time before the appearance of its first rigorous proof, in Little [25]. Since then, a number of simpler proofs have been proposed. The one in exercise 1 in section 4.1 is due to Foster [10].

The reader who wishes to know more about single-server queuing systems is advised to consult the extensive work by Cohen [4]. Two useful references on priority scheduling strategies are Conway, Maxwell and Miller [5], and Jaiswal [15]. A number of results on processor-sharing and other related scheduling strategies are presented in Coffman and Denning [3], and Kleinrock [20].

The subject of optimal scheduling, both deterministic and stochastic, has received much attention in recent years. The optimality of the dynamic rank policy was discovered independently by Gittins and Jones [12], and by Sevcik [31]. The fullest and most recent book in that area is by Whittle [34]. Discussion and proofs of conservation laws, and of several characterisation results, can be found in Gelenbe and Mitrani [11].

5
Queueing networks

Some of the most important applications of probabilistic modelling techniques are in the area of distributed computing. The term 'distributed' means, in this context, that various computational tasks that are somehow related can be carried out by different processors which may or may not be in different geographical locations. Such a broad definition covers a great multitude of systems, ranging from a few parallel processors serving a common queue of jobs, through multiprogrammed computers where one processor executes machine instructions while others perform input/output, to wide area communication networks spanning more than one continent. To study the behaviour of a distributed system, one normally needs a model involving a number of service centres, with jobs arriving and circulating among them according to some random or deterministic routing pattern. This leads in a natural way to the concept of a network of queues.

A queueing network can be thought of as a connected directed graph whose nodes represent service centres. The arcs between those nodes indicate one-step moves that jobs may make from service centre to service centre (the existence of an arc from node i to node j does not necessarily imply one from j to i). Each node has its own queue, served according to some scheduling strategy. Jobs may be of different types and may follow different routes through the network. An arc without origin leading into a node (or one without destination leading out of a node) indicates that jobs arrive into that node from outside (or depart from it and leave the network). Figure 5.1 shows a five-node network, with external arrivals into nodes 1 and 3, and external departures from nodes 1 and 4. At this level of abstraction, only the connectivity of the nodes is specified; nothing is said about their internal structure, nor about the demands that jobs place on them.

In order to define a queueing network model completely, one has to make assumptions concerning the nature of the external arrival streams, the routing of jobs among nodes and, for each node, the number of servers

Figure 5.1

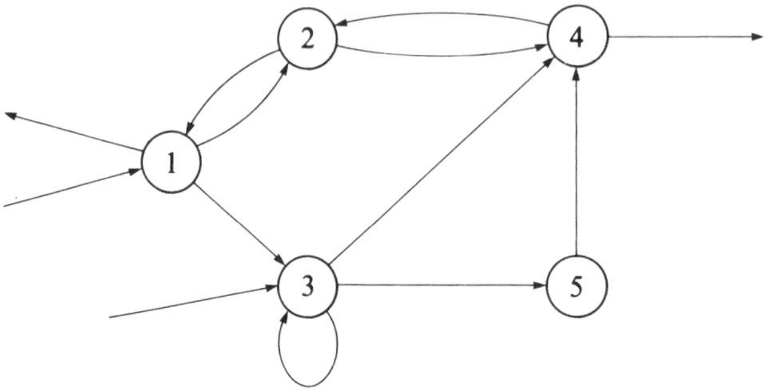

available, the required service times and the scheduling strategy. If these assumptions are sufficiently 'nice', the resulting model can be solved analytically or numerically, and the values of various performance measures can be obtained. In many other cases, where exact results are beyond our reach, good approximations can be derived by making appropriate simplifications. We shall encounter both types of models in this chapter.

There is a generally accepted classification of queueing networks, depending on (a) the topology of the underlying graph and (b) the nature of the job population. A network is said to be 'open' if there is at least one arc along which jobs enter it and at least one arc along which jobs leave it, and if from every node it is possible to follow a path leading eventually out of the network. In other words, an open network is one in which no job can be trapped and prevented from leaving. Figure 5.1 provides an illustration of an open network. By contrast, a network without external arrivals and without departures, but with a fixed number of jobs circulating forever among the nodes is called 'closed'. To turn the figure 5.1 network into a closed one, all traffic along the external incoming arcs at nodes 1 and 3, and along the outgoing ones from nodes 1 and 4, would have to be stopped.

If, at every node in the network, all jobs have the same (random) behaviour, concerning both the services they require there, and the subsequent paths they take, then we say that the network has a single job type. In a network with multiple job types, different jobs may have different service requirement characteristics and different routing patterns. Such networks can also be 'mixed', i.e. open with respect to some job types and closed with respect to others. For example, in figure 5.1 there could be two job types: one using all the arcs, and another (of which there is a fixed number of jobs) endlessly alternating between nodes 2 and 4. The network would then be mixed.

5.1   Open networks

Consider a queuing network consisting of N nodes, numbered $1, 2, \ldots, N$. Node i contains n_i identical parallel servers and, if $n_i < \infty$, an unlimited waiting room where jobs waiting for service join a single queue, in order of arrival. The required service times at node i are exponentially distributed random variables with mean $1/\mu_i$ $(i = 1, 2, \ldots, N)$.

There is a single job type. Jobs arrive from outside the network into node i in a Poisson stream with rate γ_i. Whenever a job completes service at node i, it goes to node j with probability q_{ij} $(i, j = 1, 2, \ldots, N; q_{i1} + q_{i2} + \cdots + q_{iN} \leqslant 1)$. That job leaves the network with probability q_{i0}, where

$$q_{i0} = 1 - \sum_{j=1}^{N} q_{ij}.$$

All moves are chosen without regard to past history. The probabilities q_{ij} are called the 'routing probabilities' of the network; the $N \times N$ matrix $Q = (q_{ij}); i, j = 1, 2, \ldots, N$ is its 'routing matrix'.

In this, and all subsequent queueing network models, it is assumed that the transfers of jobs from node to node are instantaneous. In cases where that assumption appears unreasonable, it is always possible to introduce transfer delays by means of artificial intermediate nodes.

Thus the network parameters are the external arrival rates, γ_i, the average service times, $1/\mu_i$, the numbers of servers at the different nodes, n_i and the routing matrix, Q. The network is assumed open. That is, at least one of the external arrival rates is non-zero and at least one of the exit probabilities q_{i0}, is non-zero (i.e. at least one of the row sums in Q is strictly less than 1). Moreover, the routing matrix is such that from every node there is a sequence of moves leading out of the network with a non-zero probability.

The above model is referred to in the literature as a 'Jackson network', after the author who first analysed it. The system state at any moment in time is described by the vector $(k_1, k_2, \ldots, k_N)$, where k_i is the number of jobs present at node i $(k_i = 0, 1, \ldots; i = 1, 2, \ldots, N)$. The assumptions of the model ensure that the process thus defined is Markov. We shall be interested in its long-term behaviour.

Suppose that the network is in the steady state. Let λ_i be the total average number of jobs arriving into (and leaving) node i per unit time. Some of the incoming jobs are external arrivals and some come from other nodes (including, perhaps, node i itself). On the average, λ_j jobs leave node j per unit time; of these, a fraction q_{ji} go to node i. Therefore, the rate of traffic from node j to node i is $\lambda_j q_{ji}$ $(j = 1, 2, \ldots, N)$. Similarly, the rate of traffic from node i to node j is $\lambda_i q_{ij}$. The job streams at node i are illustrated in figure 5.2.

Figure 5.2

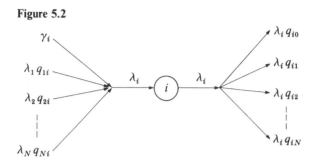

Adding together all contributions to the incoming traffic at each node, we get a set of linear equations for the unknown quantities λ_i:

$$\lambda_i = \gamma_i + \sum_{j=1}^{N} \lambda_j q_{ji}; \quad i = 1, 2, \ldots, N. \tag{1}$$

We shall call these the 'traffic equations' of the network. It can be shown that, for open networks, the traffic equations determine the arrival rates uniquely. The coefficient matrix of (1) is $I - Q$, where I is the identity matrix and Q is the routing matrix. It turns out that, from the assumption that every job in the network is able to leave it eventually, it follows that $I - Q$ is non-singular. An idea of the proof of this assertion is given in exercise 1.

It should be clear from the derivation of the traffic equations that their validity does not depend on the external arrival streams being Poisson. Only averages are involved in those equations, not distributions. Also note that the average service times play no role in (1).

Examples

1. Consider the single node Jackson network shown in figure 5.3. The external arrival stream has rate γ. After completing service, jobs re-enter the queue with probability q $(0 < q < 1)$; they leave the system with probability $1 - q$.

Figure 5.3

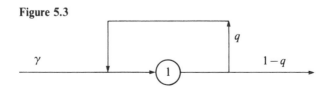

NB: the branching arcs in the figure are marked with the corresponding routing probabilities, rather than the traffic rates. This will always be our practice in future.

The routing matrix of this network consists of a single element, q. There is one traffic equation, which determines the total arrival rate, λ:

$$\lambda = \gamma + \lambda q.$$

This yields $\lambda = \gamma/(1 - q)$. The traffic rate along the feedback arc is $\lambda q = \gamma q/(1 - q)$. The rate of departures from the network is $\lambda(1 - q)$, which is of course equal to the external arrival rate, γ.

2. A university's computing service is provided by a network of N computers. Node 1 contains a large mainframe machine, while those at nodes $2, 3, \ldots, N$ are small workstations. At node i, there is an external stream of job arrivals with rate γ_i $(i = 1, 2, \ldots, N)$. A job which completes service at the mainframe computer is sent to workstation i for further processing with probability a_i $(i = 2, 3, \ldots, N)$; that job leaves the system with probability $1 - a_2 - a_3 - \cdots - a_N$. After service at workstation i, jobs go to node 1 with probability b_i and leave the system with probability $1 - b_i$ $(0 < b_i < 1)$. This network is illustrated in figure 5.4.

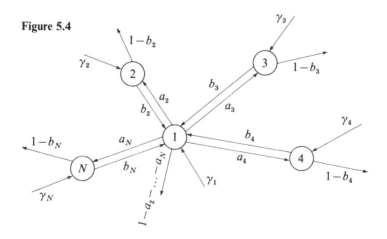

Figure 5.4

The total arrival stream at node 1 is made up of the external arrivals, plus a fraction b_j of the node j traffic $(j = 2, 3, \ldots, N)$. The stream at node i comprises the external arrivals, plus a fraction a_i of the node 1 traffic. The resulting traffic equations are as follows:

$$\lambda_1 = \gamma_1 + \sum_{j=2}^{N} \lambda_j b_j,$$

$$\lambda_i = \gamma_i + \lambda_1 a_i; \quad i = 2, 3, \ldots, N.$$

These equations are easily solved by eliminating $\lambda_2, \lambda_3, \ldots, \lambda_N$ from the first equation, determining λ_1 and hence obtaining all other arrival rates.

Let us return now to the question of existence of a steady-state distribution for a Jackson network. We have seen that, if steady-state exists, the node i is subjected to a total stream of jobs arriving at a rate λ_i, determined by the traffic equations. The load at node i (the average amount of service demanded per unit time) is $\rho_i = \lambda_i/\mu_i$. That load must be less than the number of servers available, n_i, otherwise the node would be saturated.

Thus we have an obvious necessary condition for existence of steady state: the solution of the traffic equations must be such that the resulting loads satisfy the inequalities $\rho_i < n_i$, for all $i = 1, 2, \ldots, N$.

That condition is sufficient, too. When it holds, it is possible to find a non-zero solution to the steady-state balance equations of the network, which also satisfies the normalising equation. We shall not carry out these manipulations, but will present directly the solution giving the steady-state probability, $p(k_1, k_2, \ldots, k_N)$, that there are k_1 jobs at node 1, k_2 jobs at node 2, $\ldots$, k_N jobs at node N.

Let $p_i(k)$ be the steady-state probability that there are k jobs in an isolated $M/M/n_i$ queueing system with load ρ_i. That probability is given by expressions 3.(16) and 3.(18), or by 3.(29), with n and ρ replaced by n_i and ρ_i, respectively. The steady-state distribution of the Jackson network has the form

$$p(k_1, k_2, \ldots, k_N) = p_1(k_1) p_2(k_2) \cdots p_N(k_N). \tag{2}$$

This result, known as 'Jackson's theorem', is quite remarkable. It implies that the following two statements are true.

(i) The numbers of jobs present at the various nodes are independent of each other. This conclusion is somewhat unexpected. Intuition might suggest that a long queue at a node is indicative of long queues at neighbouring nodes. However, that is not so.

(ii) Node i behaves as if it is subjected to a Poisson arrival stream with rate λ_i. This is even more surprising because, although the arrival rate into node i is λ_i, the stream of arrivals is not Poisson, in general. Even in the simple one-node network in figure 5.3, the total arrival stream (composed of the external arrivals and the feedback jobs) is not Poisson. Yet the distribution of the number of jobs present is the same as if that stream was Poisson.

The proof of the theorem amounts to demonstrating that the solution (2) satisfies the steady-state balance equations of the Markov process defined by the state vector $(k_1, k_2, \ldots, k_N)$. That demonstration will be omitted, as will the equations themselves.

Jackson's theorem allows us to derive many network performance measures by a straightforward application of existing results. To keep the expressions simple, we shall assume that all nodes contain a single server $(n_i = 1; i = 1, 2, \ldots, N)$. Every node can now be treated as an independent $M/M/1$ queue (see section 3.1). Suppose that $\rho_i < 1$ $(i = 1, 2, \ldots, N)$, so that steady state exists.

The average number of jobs at node i, L_i, is given by

$$L_i = \frac{\rho_i}{1 - \rho_i}; \quad i = 1, 2, \ldots, N. \tag{3}$$

The total average number of jobs in the network, L, is the sum of these averages over all nodes:

$$L = \sum_{i=1}^{N} L_i. \tag{4}$$

Denote by W the average time that a job spends in the network, i.e. the average interval between the arrival of a job from outside, and its departure to the outside. Applying Little's result (section 4.1) to the entire network, we get

$$L = \gamma W, \tag{5}$$

where $\gamma = \gamma_1 + \gamma_2 + \cdots + \gamma_N$ is the total external arrival rate. This, together with (4) and (3), yields

$$W = \frac{1}{\gamma} \sum_{i=1}^{N} \frac{\rho_i}{1 - \rho_i}. \tag{6}$$

One may also be interested in various other sojourns in the network. Let W_i be the average time a job spends at node i on each visit to that node. An application of Little's result to node i gives $L_i = \lambda_i W_i$, or

$$W_i = \frac{1}{\mu_i(1 - \rho_i)}; \quad i = 1, 2, \ldots, N. \tag{7}$$

Now consider the average interval, R_i, between the arrival of a job at node i and its subsequent departure from the network. Because of the memoryless routing, it does not matter whether the job arrived at node i from the outside or from another node. First, the job has to pass through node i, which takes time W_i, on the average. After that, if it goes to node j (with probability q_{ij}), its average remaining sojourn will be R_j. Hence, we can write a set of linear equations

$$R_i = W_i + \sum_{j=1}^{N} q_{ij} R_j; \quad i = 1, 2, \ldots, N. \tag{8}$$

These equations determine the conditional average sojourn times R_i uniquely, for the same reasons that the traffic equations determine the arrival rates uniquely.

What is the average number of visits that a job makes to node i during its stay in the network? That number, v_i, can be obtained by arguing as follows: on the average, a total of γ jobs enter the network from outside per unit time. Each of those jobs makes an average of v_i visits to node i. Therefore, the total average number of arrivals into node i per unit time is γv_i. On the other hand, we know that that average is λ_i. Hence,

$$v_i = \frac{\lambda_i}{\gamma}; \quad i = 1, 2, \ldots, N. \tag{9}$$

Every time a job visits node i, it requires an amount of service equal to $1/\mu_i$, on the average. The total average service, D_i, that one job requires from node i during its stay in the network, is therefore equal to

$$D_i = \frac{v_i}{\mu_i} = \frac{\rho_i}{\gamma}; \quad i = 1, 2, \ldots, N. \tag{10}$$

Rewriting this as $\rho_i = \gamma D_i$, we see that the load ρ_i can be interpreted as the average amount of work destined for node i that enters the network from outside per unit time.

When constructing a queueing network model, it is sometimes easier to estimate, or to make assumptions, about the total average service requirements, D_i, than about the requirements per visit, $1/\mu_i$, and the routing probabilities, q_{ij}. Given the quantities D_i, and the external arrival rates, one can compute the loads ρ_i without having to solve the traffic equations. This provides the average numbers of jobs at different nodes, (3), the total average number of jobs in the network, (4) and the average response time, (5). The total average time, B_i, that a job spends at node i during its life in the network can also be obtained:

$$B_i = v_i W_i = \frac{v_i}{\mu_i(1 - \rho_i)} = \frac{D_i}{1 - \rho_i}; \quad i = 1, 2, \ldots, N. \tag{11}$$

What cannot be derived from the D_i alone is the average time spent at node i per visit, W_i, and the conditional average sojourn time in the network R_i.

The results that we have presented can easily be re-derived for networks where there are more than one servers per node. If, for example, node i contains an unlimited number of parallel servers ($n = \infty$), then the $M/M/\infty$ formulae should be applied to it: the average number of jobs present is $L_i = \rho_i$, the average time a job spends in it per visit is $W_i = 1/\mu_i$ and the total average time a job spends in it is $B_i = D_i = v_i/\mu_i$. For such a node, it is not even necessary that the service times are exponentially distributed; they may be general.

Thus, Jackson's theorem gives us the steady-state distribution of the number of jobs in the network. In conjunction with Little's result, it also provides average sojourn times, both at a node and in the network. The latter may be unconditional, or conditioned upon the starting node. It should be emphasised, however, that the theorem does not say anything about the distribution functions of the sojourn times, either at a given node, or in the network. In particular, it is not clear *a priori* whether the distribution of the sojourn time at node i is the same as the distribution of the response time in an isolated $M/M/n_i$ system with parameters λ_i and μ_i.

The last statement turns out to be true. It can be shown that, despite the fact that the total arrival stream into node i is not necessarily Poisson, the arriving jobs can still be treated as random observers. In other words, a job arriving into node i, whether from outside or from another node, will see the steady-state $M/M/n_i$ distribution of the number of jobs present. This implies, for instance, that if node i contains a single server, then the distribution function of the sojourn times there, $F_i(x)$, is exponential, with parameter $\mu_i - \lambda_i$ (section 3.1):

$$F_i(x) = 1 - e^{-(\mu_i - \lambda_i)x}; \quad x \geqslant 0. \tag{12}$$

The problem concerning the distribution of sojourn times in the whole network, or along paths in the network consisting of several nodes, is considerably more complex. The reason for this is that there may be dependencies between the sojourn times at different nodes. To illustrate the phenomena involved, consider the simple three-node network in figure 5.5.

Figure 5.5

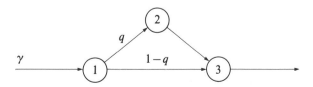

This is an example of a 'feedforward network', i.e. one where a job can never visit the same node more than once. There is only one point where routing decisions are taken: after completing service at node 1, jobs go to node 2 with probability q and to node 3 with probability $1 - q$. The (external) arrival rate at node 1 is γ. It is readily seen that the steady-state arrival rates at nodes 2 and 3 are $q\gamma$ and γ, respectively. For non-saturation, those rates have to satisfy $\gamma < \mu_1$, $q\gamma < \mu_2$ and $\gamma < \mu_3$. Assume that these inequalities hold and that the network is in steady-state.

Because there is no feedback, and because of Burke's theorem (section 3.2), all traffic streams in this network are Poisson. Indeed, since only external jobs arrive into node 1, that stream is Poisson. Hence, the departures from node 1 are Poisson and remain so after splitting in two. Since the arrivals into node 2 are Poisson, so are the departures from node 2. The merger of two Poisson streams at node 3 is also Poisson, and so are the departures from node 3. Another consequence of Burke's theorem is that, when a job leaves node 1 (or 2) to go to node 2 (or 3), then the numbers of jobs, and the sojourn times, at the node left and the node joined are independent of each other.

The sojourn time of a job in the network is the sum of its sojourn times at the nodes through which it passes. Denote the latter by T_1, T_2 and T_3 (T_2 may or may not be included in the sum). We know that these three random variables are exponentially distributed, with parameters $\mu_1 - \gamma$, $\mu_2 - q\gamma$ and $\mu_3 - \gamma$, respectively. If a job takes the path $(1, 3)$, then, in view of the above remarks, T_1 and T_3 are independent of each other and the total sojourn time distribution is the convolution of two exponential distributions.

Suppose now that the job takes the path $(1, 2, 3)$. The times T_1 and T_2 are still independent of each other, as are also T_2 and T_3. However, T_1 and T_3 are dependent random variables. If T_1 is large then, when the job leaves node 1, there is likely to be a long queue left behind. While our job passes through node 2, many of the jobs from that long queue will overtake it by taking the direct path $(1, 3)$ and will delay it at node 3, causing T_3 to be large. This dependency between T_1 and T_3 makes the problem of finding the distribution of $T_1 + T_2 + T_3$ extremely difficult. An analytical solution exists for the special case that we are discussing now, but in more general networks where overtaking is possible the problem is still open.

Note that the dependency between T_1 and T_3 does not contradict the fact that, both when a job leaves node 1 and at a random observation point, the numbers of jobs at nodes 1 and 3 are independent of each other. The sojourn times' dependency arises because, rather than taking a snapshot of the system, we are following the progress of a job through the network. That makes it possible for events in the past to influence the future.

To recapitulate, the average sojourn times along arbitrary paths in arbitrary Jackson networks are easy to obtain (the mean of a sum is equal to the sum of means, regardless of whether the random variables are dependent or not). The distributions of such sojourn times are unattainable, except in some special cases.

Exercises

1. Consider the traffic equations (1), with $\gamma_i = 0$ for all $i = 1, 2, \ldots, N$. If the matrix $I - Q$ was singular then those equations would have a non-zero solution. That would imply that the long run total arrival rates need not be all zero, even if the external ones are. Show that this contradicts the assumption that the network is open. Hence, $I - Q$ must be non-singular.

2. For the network in example 2 in this section, assume that at node i there is a single exponential server with average service times $1/\mu_i$. Find the average time that the jobs submitted at node i spend in the network.

3. In a three-node Jackson network, the flow of jobs is governed by a routing matrix, $Q = (q_{ij})$ $(i, j = 1, 2, 3)$, given by

$$Q = \begin{bmatrix} 0 & 0.2 & 0.8 \\ 0 & 0 & 0 \\ 1 & 0 & 0 \end{bmatrix}.$$

There are external arrivals at nodes 1 and 2, with rates 5 and 3 jobs per second, respectively. The average service times at nodes 1, 2 and 3 are 30 milliseconds, 100 milliseconds and 40 milliseconds, respectively. There is a single server at each node.

Find the average numbers of jobs at the three nodes and in the whole network. Also determine the average sojourn times in the network for the jobs submitted at nodes 1 and 2.

4. For an arbitrary Jackson network, let v_{ij} be the average number of visits that a job which is about to join node i will make to node j before leaving the network. Show that these quantities satisfy the following set of equations:

$$v_{ii} = 1 + \sum_{k=1}^{N} q_{ik} v_{kj}; \quad i = 1, 2, \ldots, N.$$

$$v_{ij} = \sum_{k=1}^{N} q_{ik} v_{kj}; \quad i \neq j; i, j = 1, 2, \ldots, N.$$

Use those equations in the context of the network in the previous exercise and find the average numbers of visits that a job submitted at node 1 makes to node 1 and to node 3. Compare the results with the averages v_1 and v_3, obtained according to (9), and explain the discrepancy.

5.2   Closed networks

There are many systems where jobs can be delayed at a number of nodes, and where the total number of jobs available, or admitted, is kept constant over long periods of time. Such systems are modelled by closed queuing networks, i.e. ones without external arrivals and without departures. We have already encountered examples of that type in chapter 3. The terminal system models of section 3.3 (figures 3.7 and 3.9) can be regarded as closed networks with two nodes: one containing K terminals and one containing a processor (or n processors). A fixed number, K, of jobs circulate among the two nodes at all times.

Here we shall consider closed queuing networks with N nodes, each of which contains either a single server, or an unlimited number of servers (at least as many as there are jobs). The latter type of nodes are referred to as 'delay nodes', to emphasise the fact that jobs do not queue there; they are just delayed independently of each other. A delay node can be used to model a collection of terminals.

The reason for restricting the discussion to single-server and delay nodes is simplicity. Various generalisations will be mentioned later.

The average service time at node i is $1/\mu_i$ $(i = 1, 2, \ldots, N)$. The distribution of service times at a single-server node is assumed to be exponential, but at a delay node it may be general (at those nodes, we may talk about 'delay times' or 'think times', instead of service times). After completing service at node i, a job goes to node j with probability q_{ij} $(i, j = 1, 2, \ldots, N)$. These routing probabilities now satisfy

$$\sum_{j=1}^{N} q_{ij} = 1; \quad i = 1, 2, \ldots, N. \tag{13}$$

Thus, no job ever leaves the network. There are no external arrivals, either. Hence, the number of jobs in the network is always constant, and is equal to the number that were there at time 0. Denote that number by K. The state of the network at any time is described by the vector $(k_1, k_2, \ldots, k_N)$, where k_i is the number of jobs at node i. Since the only feasible states are those for which

$$\sum_{i=1}^{N} k_i = K, \tag{14}$$

the state space, S, is obviously finite. More precisely, the number of states in S is equal to the number of ways in which the integer K can be partitioned into N non-negative components. That number is easily seen (exercise 1) to be equal to

$$|S| = \binom{K + N - 1}{N - 1}. \tag{15}$$

Therefore, the network always reaches steady-state, no matter what the values of its parameters. Assume that it has done so.

Denote by λ_i the total average number of jobs arriving into node i per unit time ($i = 1, 2, \ldots, N$). These jobs can only come from other nodes in the network. Following the same reasoning as in the case of an open network, we can write a set of traffic equations that the arrival rates must satisfy:

$$\lambda_i = \sum_{j=1}^{N} \lambda_j q_{ji}; \quad i = 1, 2, \ldots, N. \tag{16}$$

These equations are now homogeneous, because of the absence of external arrival rates. Moreover, the coefficient matrix, $I - Q$, is singular, since all its row sums are 0. Hence, the set (16) has infinitely many solutions, which differ from each other by multiplicative constants.

Let $(e_1, e_2, \ldots, e_N)$ be any non-zero solution to the traffic equations. Such a solution can be obtained by fixing one of the e_is arbitrarily and then using the equations to determine the others. The quantities e_i, if not equal to the arrival rates λ_i, are proportional to them: $e_i = c\lambda_i$ ($i = 1, 2, \ldots, N$), for some non-zero constant, c. Similarly, the ratios $\sigma_i = e_i/\mu_i$ are proportional to the loads $\rho_i = \lambda_i/\mu_i$.

The steady-state distribution of the number of jobs in the network, $p(k_1, k_2, \ldots, k_N)$, turns out to have a product form, similar to the one specified by Jackson's theorem for open networks. The idea is to treat node i as an isolated $M/M/1$ or $M/M/\infty$ queuing system, depending on whether it is a single-server or a delay node. Assuming that such a system is subjected to a load $\sigma_i = e_i/\mu_i$, and applying the appropriate formula for the probability that there are k jobs present, would yield some value, $r_i(k)$. It can be shown that the probability $p(k_1, k_2, \ldots, k_N)$ is proportional to the product of the quantities $r_i(k_i)$, taken over the N nodes:

$$p(k_1, k_2, \ldots, k_N) = \frac{1}{G} r_1(k_1) r_2(k_2) \cdots r_N(k_N). \tag{17}$$

The network state $s = (k_1, k_2, \ldots, k_N)$ must of course belong to the set of feasible states, S. That is, relation (14) must hold. The constant G is a normalising one. It is determined by the condition that the sum of the probabilities (17), over all $s \in S$, is 1. This yields,

$$G = \sum_{s \in S} r_1(k_1) r_2(k_2) \cdots r_N(k_N).$$ (18)

There are obvious similarities between the above result and Jackson's theorem, but also some important differences which should be pointed out. First, the fact that (14) holds implies that the numbers of jobs present at the N nodes are not independent random variables (e.g. if there are K jobs at node 1, then all other nodes must be empty). Also, the factor $r_i(k_i)$ which appears in the right-hand side of (17), is not necessarily equal to the probability that there are k_i jobs at node i. The M/M/1 or M/M/∞ expression used in deriving $r_i(k_i)$, is a formal device only. It does not really describe the behaviour of node i. Finally, if $r_i(k_i)$ is replaced, for all $i = 1, 2, \ldots, N$, by

$$d_i c^{k_i} r_i(k_i),$$

where d_i and c are arbitrary non-zero constants, then the only thing that will change in (17) is the normalisation constant, G. If the latter is re-computed appropriately, along the lines of (18), then the probabilities given by (17) will have the same values. It is because of this last property that, no matter which solution of (16) is used, (17) and (18) yield the correct steady-state distribution.

The product form solution is proved by demonstrating that the expressions (17) satisfy the steady-state balance equations of the Markov process whose state is the vector $(k_1, k_2, \ldots, k_N)$. We shall not do this.

From a computational point of view, there is exactly one difficult step in implementing the solution suggested by (16) and (18), and hence in obtaining various network performance measures. That step consists of determining the normalisation constant, G. The reason for the difficulty is of course the number of terms involved in the summation (18). That number, given by (15), increases rather quickly with N and K. Several algorithms exist for computing G efficiently (one of them is outlined in exercise 2). However, we shall not pursue that avenue any further because most of the performance measures that are of interest to us can be obtained in a simpler and more intuitively appealing way.

The approach that we are about to describe is called 'mean value analysis'. It is based on nothing more complicated than Little result. The fundamental quantities of interest are the average number of jobs at node i, L_i, and the average time a job spends at node i (on each visit to that node), W_i. In addition, we shall wish to talk about the average time a job spends in the network, W, the average number of visits that a job makes to node i, v_i, and the throughput, T (i.e. the average number of jobs leaving the network per unit time). Obviously, in order to do this in the context of a closed network, it is necessary to re-introduce external arrivals and departures, without otherwise disturbing the model.

In a real system where the number of jobs is kept constant, the membership of the set of jobs may change. When a job is completed, it departs and is immediately replaced by a new one from outside. This is what will happen in our model. Assume that there is a special point on one of the arcs in the network, called the 'entry point' and denoted by 0. Whenever a job traverses that particular arc and passes through point 0, it changes into a new job. That is, the old job departs from the network, a new job enters it immediately and goes to the destination node of the arc. The exchange is illustrated in figure 5.6.

Figure 5.6

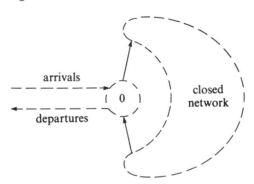

Clearly, this modification of the model has no effect other than to transform the jobs from permanent entities into temporary ones. What was before the period between two consecutive passes of the same job through point 0 is now the sojourn time of a job in the network. Consequently, the performance measures W, v_i and T acquire meaning and interest.

Since an average of T jobs arrive into the network per unit time, and each of them makes an average of v_i visits to node i, the total arrival rate into node i is

$$\lambda_i = Tv_i; \quad i = 1, 2, \ldots, N. \tag{19}$$

These relations imply that the averages v_i satisfy the traffic equations (16): a fact which could also have been established directly. If one of the v_is is known, all the others are determined uniquely by those equations. But we do know that each job makes exactly one passage through point 0 during its sojourn in the network. Suppose that the arc which contains the entry point leads from node a to node b. Then each visit to node a is followed by a passage through point 0 with probability q_{ab}. Hence, $v_a q_{ab} = 1$, or

$$v_a = \frac{1}{q_{ab}}.$$

This together with (16), determines all v_i $(i = 1, 2, \ldots, N)$.

An application of Little's result to node i yields a relation between the average number of jobs at that node, L_i, and the average time jobs spend there on each visit, W_i:

$$L_i = \lambda_i W_i = T v_i W_i; \quad i = 1, 2, \ldots, N. \tag{20}$$

Summing these over all nodes and remembering that the total number of jobs in the network is K, we get

$$K = T \sum_{i=1}^{N} v_i W_i,$$

or

$$T = \frac{K}{\sum_{i=1}^{N} v_i W_i}. \tag{21}$$

The average sojourn time at node i, W_i, depends on the type of the node and on the number of jobs found there by an incoming job. If node i is a delay node, there is no queue and W_i is equal to the average service time, $1/\mu_i$. If node i is a single-server node, and if when a job gets there it finds an average of Y_i jobs already present, then W_i is equal to $Y_i + 1$ average service times. Introducing an indicator, δ_i, which has value 0 if node i is a delay node and 1 if it is a single-server node, we can write

$$W_i = \frac{1}{\mu_i} (1 + \delta_i Y_i); \quad i = 1, 2, \ldots, N. \tag{22}$$

Note that the state of node i seen by an incoming job has a different distribution from the state seen by a random observer. For instance, an incoming job can never see all K jobs at node i, because it itself cannot be among the jobs already present. That is why Y_i is different from L_i. Yet if one could somehow relate these two types of averages, all performance measures would be determined from equations (20), (21) and (22).

A way out of the difficulty is provided by the following result, which will be referred to as the 'job observer property': the network state seen by a job in transit from one node to another (that job is considered to have left the former node and not to have joined the latter), has the same distribution as the state which a random observer would see if the total number of jobs circulating in the network was not K, but $K - 1$. In other words, a job about to join a node behaves as a random observer of a network whose job population is smaller by one. This is quite an intuitive assertion, in view of the fact that a job in transit cannot see itself at any node. However, the formal proof is non-trivial and will be omitted.

We can now develop a solution procedure based on a recurrence with respect to the job population size, K. The quantities T, W_i, L_i and Y_i will be regarded as functions of K (the visit averages, v_i, depend only on the routing matrix and not on the number of jobs). The job observer property implies that

$$Y_i(K) = L_i(K - 1); \quad i = 1, 2, \ldots, N. \tag{23}$$

This allows us to rewrite equations (22), (21) and (20) in the form of recurrence relations in K:

$$W_i(K) = \frac{1}{\mu_i} \left[1 + \delta_i L_i(K - 1) \right]; \quad i = 1, 2, \ldots, N. \tag{24}$$

$$T(K) = \frac{K}{\sum_{i=1}^{N} v_i W_i(K)}. \tag{25}$$

$$L_i(K) = v_i T(K) W_i(K); \quad i = 1, 2, \ldots, N. \tag{26}$$

The obvious starting condition for an iterative procedure based on these equations is $L_i(0) = 0$ $(i = 1, 2, \ldots, N)$. The first application of the procedure would yield $W_i(1)$, $T(1)$ and $L_i(1)$; the second gives $W_i(2)$, $T(2)$ and $L_i(2)$, and so on, until the desired population size is reached.

The total average time that a job spends at node i during its life in the network, $B_i(K)$, is equal to the average sojourn time at node i multiplied by the average number of visits to that node:

$$B_i(K) = v_i W_i(K); \quad i = 1, 2, \ldots, N. \tag{27}$$

Similarly, the total average service time that a job requires from node i, D_i (that quantity does not depend on K), is equal to

$$D_i = v_i \frac{1}{\mu_i}; \quad i = 1, 2, \ldots, N. \tag{28}$$

Sometimes the total average service requirements D_i can be estimated quite easily from measurements. They may therefore be given as the network parameters, instead of the individual average service times $1/\mu_i$ and the routing matrix Q. In that case, the recurrence schema (24), (25), (26) is replaced by a simpler one which does not involve the visit averages v_i. The role of $W_i(K)$ is played by $B_i(K)$ and that of $1/\mu_i$ by D_i:

$$B_i(K) = D_i[1 + \delta_i L_i(K - 1)]; \quad i = 1, 2, \ldots, N. \tag{29}$$

$$T(K) = \frac{K}{\sum_{i=1}^{N} B_i(K)}. \tag{30}$$

$$L_i(K) = T(K)B_i(K); \quad i = 1, 2, \ldots, N. \tag{31}$$

The total average time that a job spends in the network, $W(K)$, is given by

$$W(K) = \sum_{i=1}^{N} v_i W_i(K) = \sum_{i=1}^{N} B_i(K) = \frac{K}{T(K)}. \tag{32}$$

The last equality follows also from Little's result, applied to the entire network.

Another performance measure which is often of interest in connection with a single-server node, i, is its utilisation, $U_i(K)$ (i.e. the fraction of time that the server is busy). As in all single-server queueing systems, $U_i(K)$ is equal to the load on node i:

$$U_i(K) = \frac{\lambda_i}{\mu_i} = T(K)\frac{v_i}{\mu_i} = T(K)D_i. \tag{33}$$

This result provides us with a simple upper bound on the network throughput. Since the utilisation of a server cannot exceed 1, we must have

$$T(K) \leqslant \min_i \frac{1}{D_i}, \tag{34}$$

where the minimum is taken over all single-server nodes.

If node i is a delay node, then the right-hand side of (33) is equal to the average number of jobs, or busy servers there.

The solution procedures that we have described evaluate $2N + 1$ quantities (W_i, L_i and T, or B_i, L_i and T), for each population size up to the desired one. The necessity of stepping through all the intermediate values of K can be avoided if one is willing to accept an approximate solution instead of an exact one. The idea is to relate the job-observed and random-observed node averages, Y_i and L_i, at the same population level. One such relation that has gained acceptance is

$$Y_i(K) = \frac{K - 1}{K} L_i(K); \quad i = 1, 2, \ldots, N. \tag{35}$$

This expression is of course exact for $K = 1$. It is also asymptotically exact for large values of K. In general, experience indicates that the approximations obtained with the aid of (35) are sufficiently accurate for most practical purposes.

The use of (35) reduces the relations (24), (25) and (26), or (29), (30) and

(31), to fixed-point equations expressing the $2N + 1$ unknowns in terms of themselves. All but one of the unknowns can be eliminated explicitly. For instance, replacing $L_i(K - 1)$ in (29) with the right-hand side of (35) and using (31), we get

$$B_i(K) = \frac{D_i}{1 - \dfrac{K - 1}{K} \delta_i D_i T(K)}; \quad i = 1, 2, \ldots, N. \tag{36}$$

This allows $B_i(K)$ to be eliminated from (30), turning the latter into an equation for the throughput alone, of the form

$$T(K) = f[T(K)].$$

There are many iterative methods for solving such equations.

Example

A multiprogrammed computer under heavy demand can be modelled by a closed network of the type illustrated in figure 5.7.

Figure 5.7

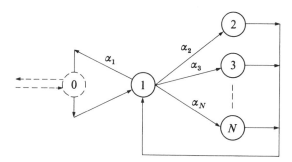

Node 1 represents the central processing unit, while nodes $2, 3, \ldots, N$ are various input/output devices. After completing a service at the CPU, a job goes to device i with probability α_i ($i = 2, 3, \ldots, N$); with probability $\alpha_1 = 1 - \alpha_2 - \cdots - \alpha_N$, the job leaves the network and is immediately replaced at the CPU by a new job. This last move takes place along the feedback arc from node 1 to node 1, which contains the entry point, 0. All nodes are single-server ones, with exponentially distributed service times (mean $1/\mu_i$ at node i) and FIFO queueing. The number of jobs in the network, i.e. the degree of multiprogramming, is K.

The traffic equations for this network, satisfied by the average numbers of visits that a job makes to the different nodes, are as follows:

$$v_1 = \alpha_1 v_1 + \sum_{j=2}^{N} v_j,$$

$$v_i = \alpha_i v_1; \quad i = 2, 3, \ldots, N.$$

In addition, the fact that the entry point is on the feedback arc implies that $v_1 = 1/\alpha_1$. This yields $v_i = \alpha_i/\alpha_1$, for $i = 2, 3, \ldots, N$.

The total average amounts of service that a job requires from the CPU and the I/O devices are

$$D_1 = \frac{1}{\alpha_1 \mu_1}; \quad D_i = \frac{\alpha_i}{\alpha_1 \mu_i}; \quad i = 2, 3, \ldots, N.$$

These total service requirements may in some cases be given as the network parameters, instead of α_i and $1/\mu_i$.

Perhaps the most important performance measure for this system is its throughput, $T(K)$. It is determined by means of the recurrence relations. Having obtained $T(K)$, the average job response time is given by (32) and the utilisations of the various nodes by (33).

Exercises

1. The state of a closed network with N nodes and K jobs (or a partition of the integer K into N components) can be represented by a string of N zeros and K ones. Each node, starting with the first, contributes to the string a 0 followed by as many 1s as there are jobs present at that node. Thus, for a network with three nodes and five jobs, the string 01001111 represents the state where there is one job at the first node, no jobs at the second and four jobs at the third.

Using that representation, prove (15) by arguing that the number of possible states is equal to the number of ways of choosing the positions of $N - 1$ zeros among $K + N - 1$ digits.

2. The normalisation constant of a closed network, given by (18), depends on the number of nodes and on the number of jobs. To make that dependence explicit, denote the constant by $G_N(K)$.

Suppose that all nodes in the network are single-server ones. Show that the factors $r_i(k_i)$, appearing in (17), can be taken as σ^{k_i}, where $\sigma_i = e_i/\mu_i$ and e_i is any solution of (16). Expression (18) then becomes

$$G_N(K) = \sum_{s \in S} \sigma_1^{k_1} \sigma_2^{k_2} \cdots \sigma_N^{k_N}.$$

Split this sum, which extends over all possible states $s = (k_1, k_2, \ldots, k_N)$, into two sums: the first extends over those states for which $k_N = 0$ and the second over the states for which $k_N > 0$. Demonstrate that the resulting expression can be written as

$$G_N(K) = G_{N-1}(K) + \sigma_N G_N(K-1).$$

This recurrence suggests an efficient algorithm for computing $G_N(K)$, starting with the initial conditions $G_0(K) = 0$ $(K = 1, 2, \ldots)$ and $G_N(0) = 1$ $(N = 1, 2, \ldots)$.

3. Implement the solution procedure for the multiprogrammed computer example in this section. In the special case of $N = 2$, compare the answers obtained for the throughput and the average response time with those arrived at by a different method in exercise 1, section 3.3.

4. Use the program from the previous exercise, appropriately modified, if necessary, to evaluate the performance of a closed network consisting of N single-server nodes which are visited by the jobs in strict rotation: the moves are from node i to node $(i \bmod N) + 1$ $(i = 1, 2, \ldots, N)$. The entry point is on the arc from node N to node 1. The average service time at node i is $1/\mu_i$.

5.3   Optimal multiprogramming

In a real-life multiprogrammed computer system, the behaviour of the active jobs (i.e. the ones that are currently being executed), is influenced by their number. The main reason for this is that the total demand for main memory may easily exceed the amount of main memory available. When that happens, the operating system must take some action in order to enable all active jobs to continue execution. One possibility is to swap pages of information between the main and peripheral memories. Alternatively, whole jobs may be swapped in and out of main memory. In both cases, the higher the degree of multiprogramming, the greater the input/output traffic generated as a result of memory contention. In the limit, that traffic becomes the dominant activity in the system: all jobs spend most of their time waiting for the memory they need to become available to them and hardly any useful work is done. The throughput approaches 0 and the response time tends to infinity. This phenomenon is known as 'thrashing'.

To avoid thrashing, the level of multiprogramming should be controlled and prevented from becoming too large. It is not desirable to let it drop too low, either, because then the computing resources are under-utilised and the throughput is less than the maximum that can be attained. For any system configuration, there is a degree of multiprogramming which is optimal with respect to the average number of job completions per unit time. In this section, we shall tackle the problem of discovering and achieving that optimum.

Our main modelling tool will be the closed network introduced in connection with the multiprogrammed computer example in the previous section. That network is shown in figure 5.7. It is not, of course, the most realistic model that could have been adopted, but it is sufficient for our purposes.

The system consists of one CPU and $N - 1$ I/O devices ($N \geqslant 2$), and the degree of multiprogramming is K. We shall now introduce a new feature, namely a main memory of size M pages. The K jobs in the system have to share this memory. Assume that one of the I/O devices – say node 2 – is used for the service of I/O requests arising from memory contention. Regardless of whether those requests involve swaps of individual pages, or of memory segments corresponding to entire jobs, we shall refer to them as 'page faults'.

In the absence of memory contention, the life of a job in the system would consist of one or more visits to the CPU, alternating with visits to I/O devices. Let the total average service requirement from node i be D_i ($i = 1, 2, \ldots, N$). These requirements are given as model parameters and are independent of K and M. However, the competition for main memory causes the CPU services of a job to be interrupted from time to time by page faults. When a page fault occurs, the interrupted job has to leave node 1 and request a service from node 2.

The paging behaviour of a job is governed by what is called its 'lifetime function', $h(x)$. Given the amount of main memory, x, available to the job, the lifetime function specifies the average CPU interval between consecutive page faults. Clearly, $h(x)$ is an increasing function of x: the more memory a job has at its disposal, the less frequently it experiences page faults. The shape of the lifetime function is a job characteristic which, along with the other job characteristics, has to be established or assumed as part of the model construction. Two examples of lifetime functions that have been used in the literature are illustrated in figure 5.8. Both examples display a 'flattening out' of the function for large values of x. This is due to the fact that, after a certain point, giving more memory to a job ceases to have an effect on its page fault rate.

Figure 5.8

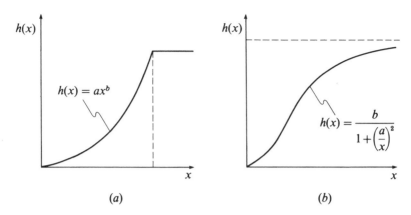

(a)   (b)

With K jobs sharing M pages of main memory, it is reasonable to suppose that the amount of memory available to one job is M/K. Hence, the average amount of CPU service that a job receives between consecutive page faults is $h(M/K)$ (that service need not be contiguous; it may be interrupted by visits to I/O devices for reasons other than memory contention). Since the total average CPU service required by a job is D_1, the average number of page faults that a job experiences during its life in the system is $D_1/h(M/K)$. Each of those page faults results in a visit to node 2, where the average service time is $1/\mu_2$. Denote by $H(K)$ the total average service that a job requires from node 2 as a result of the competition for main memory (this is of course in addition to the requirement D_2, which is inherent to the job and does not depend on K). The above remarks imply that

$$H(K) = \frac{D_1}{\mu_2 h\left(\dfrac{M}{K}\right)}; \quad K = 1, 2, \ldots . \tag{37}$$

Note the behaviour of this quantity as the degree of multiprogramming increases. When $K \to \infty$, the ratio M/K tends to 0 and so does $h(M/K)$. The service requirement $H(K)$ approaches infinity. This, together with the bound (34), shows very clearly that the system throughput drops to 0, thus providing a more formal explanation of the threshing phenomenon.

It is of interest to plot the value of the throughput, $T(K)$, against the degree of multiprogramming, K. To evaluate $T(K)$, one could use either the exact recurrence equations, or the approximate fixed-point ones, of the mean value analysis (see section 5.2). If the former approach is adopted, it should be remembered that while the recurrence with respect to the population size is in progress, the service requirement $H(K)$ remains constant. It changes, however, with the degree of multiprogramming. To make this absolutely clear, let us use a different symbol, k, to denote an intermediate population size; K will still be the desired degree of multiprogramming. Equations (29), (30) and (31) can now be rewritten, bearing in mind the peculiarity of node 2, as follows:

$$B_2(k) = [D_2 + H(K)][1 + L_2(k-1)]; \quad k = 1, 2, \ldots, K.$$
$$B_i(k) = D_i[1 + L_i(k-1)]; \quad i \neq 2; k = 1, 2, \ldots, K.$$

(38)

$$T(k) = \frac{k}{\sum_{i=1}^{N} B_i(k)}; \quad k = 1, 2, \ldots, K. \tag{39}$$

$$L_i(k) = T(k) B_i(k); \quad i = 1, 2, \ldots, N; k = 1, 2, \ldots, K. \tag{40}$$

Thus the computation would involve an inner loop with respect to k and an outer one with respect to K, with re-evaluation of $H(K)$ in the outer loop.

A typical plot of $T(K)$ is shown in figure 5.9(a).

Figure 5.9

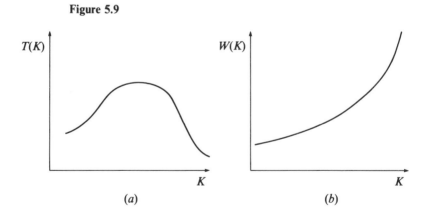

(a)   (b)

What is usually observed is that initially the throughput increases with the degree of multiprogramming. That is the region where the system is on the whole under-utilised, and adding more jobs helps to utilise it better. Then there is often a plateau where for a while the addition of further jobs does not have much effect. After that, the throughput goes into a steep decline, signalling the onset of thrashing. The aim of any policy for controlling the degree of multiprogramming should be to maintain the latter in the plateau region, at or near the optimal point.

The average response time is related to the throughput in accordance with (32). A typical plot of $W(K)$ looks like the one in figure 5.9(b).

We shall now modify the model by including in it some sources of demand. Instead of K jobs circulating around the N nodes at all times, assume that there are K statistically identical terminals from which jobs are being submitted from time to time. When a terminal submits a job for execution, the latter joins the CPU node; it then circulates as before between the CPU and I/O devices, eventually exiting from the CPU back to the terminal from which it came. That terminal goes into a think state for a random period with mean $1/\gamma$, after which it submits a new job, etc. The modified model is shown in figure 5.10.

Figure 5.10

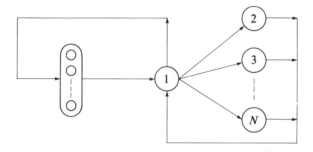

We shall continue to think of the collection of CPU, I/O devices and M pages of main memory as 'the computer'. The number of active jobs, j, or the degree of multiprogramming, is no longer constant but may vary in the range $j = 0, 1, \ldots, K$. When j jobs are active, there are $K - j$ terminals in think state, or $K - j$ passive jobs at the terminals. The assumptions concerning the total average service requirements per job, and the role of the lifetime function, are the same as before.

The performance measures of interest are the throughput, T, and the response time, W, or the average number of job completions per unit time and the average interval between a job submission and its completion, respectively. Being able to calculate these quantities for different values of K would enable one to examine the effect which an increase in the number of terminals has on the performance of the system.

At first glance, it appears that here we have a closed queuing network of the same type as the one in figure 5.7. All that has happened is that the entry point has been replaced by a delay node representing the terminals. It would seem that the new model can be solved in the same way as the old one. That, however, is not the case. An essential feature of the present network is that certain patterns of job behaviour, such as the frequency with which jobs visit node 2, depend on the network state. More precisely, they depend on the number of active jobs. This property, which is due to the fact that only the active jobs share the main memory, prevents the application of the standard solution methods and necessitates a different approach.

We shall employ an approximation called 'decomposition'. This consists of two steps. First, the sub-network representing the computer is treated as an isolated closed network, as has been done already, and is solved for all degrees of multiprogramming in the range $j = 1, 2, \ldots, K$. This yields, for each j, a corresponding value of the throughput, $T(j)$.

The second step is to replace the entire computer sub-network by one single-server node which, when there are j jobs present, completes them at the rate of $T(j)$ per unit time. The resulting two-node network is shown in figure 5.11.

Figure 5.11

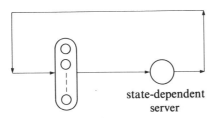

state-dependent
server

Now the system can be modelled by a simple Markov process whose state, j, is the number of jobs at the computer node. The possible states are the integers $\{0, 1, \ldots, K\}$. This model bears a very close resemblance to the terminal system one in section 3.3. The state diagram of the present process differs from the one in figure 3.8 only in its state-dependent transition rates from state j to state $j - 1$ (see figure 5.12).

Figure 5.12

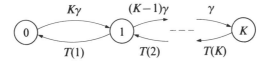

Denoting by p_j the steady-state probability that the process is in state j, we can write the following set of balance equations:

$$(K - j + 1)\gamma p_{j-1} = T(j)p_j; \quad j = 1, 2, \ldots, K. \tag{41}$$

These equations, together with the normalising one, are easily solved for the probabilities p_j. From that solution, the system throughput is obtained as a weighted average:

$$T = \sum_{j=1}^{K} p_j T(j). \tag{42}$$

According to Little's result, the average number of jobs at the computer node is TW. Similarly, the average number of jobs at the terminals is T/γ. Since the sum of these two averages is K, we can write an equation which determines the average response time, W:

$$K = T\left(W + \frac{1}{\gamma}\right) \tag{43}$$

(see also equation 3.(47)).

The second step in the decomposition approximation amounts, in essence, to assuming that the computer sub-network reaches steady-state in between successive changes of the degree of multiprogramming. In practical applications, this is justified by the fact that jobs circulate much faster within the sub-network than between it and the terminals (e.g. the intervals between consecutive page faults and the I/O service times are usually measured in milliseconds, while think times and total service requirements are measured in seconds). It is thus acceptable to regard the computer sub-network as immediately adopting a new steady-state behaviour every time a job enters it or leaves it.

If T and W are plotted against the number of terminals, K, the resulting graphs look much like the ones in figure 5.9. The occurrence of thrashing may be delayed if the think times are large, but it cannot be avoided.

We conclude that, in all systems where the main memory is a critical resource, it is necessary to have some policy for controlling the degree of multiprogramming. The main difficulty in devising such a policy to optimal effect is that the lifetime function of the jobs submitted for execution is not usually known in advance. One way of dealing with this is to observe the behaviour of the jobs in the system and maintain an empirical estimate of their lifetime function. Another possibility is to use an indirect control criterion based on some symptom of thrashing.

Consider, for example, the activity of the paging device (node 2 in our model). We have seen that the thrashing phenomenon is accompanied by high page fault rates and hence high level of demand for paging services. The utilisation of the paging device is therefore a good indicator of thrashing. A control algorithm based on that indicator would monitor the paging device and keep a running estimate of the fraction of time that it is busy. As soon as that fraction exceeds a certain threshold – say 50 % – new jobs are denied admission into the computer sub-network.

Exercise

1. Implement the solution of the terminal system model described in this section. Assume that the computer sub-network consists of one CPU and two I/O devices, of which the first is the paging device. The total average service requests (inherent to the jobs) directed at the three nodes are $D_1 = 5$ seconds, $D_2 = 0$ seconds and $D_3 = 8$ seconds. The lifetime function is of the form displayed in figure 5.8(a):

$$h(x) = \begin{cases} ax^{1.5}, & \text{for } 0 \leqslant x \leqslant 100 \\ a100^{1.5}, & \text{for } x \geqslant 100 \end{cases}.$$

Vary the coefficient a in the range $[10^{-6}, 10^{-3}]$, the main memory size in the range $[100, 1000]$ pages and the average think times in the range $[0.1, 2]$ seconds. In each case, plot the system throughput against the number of terminals and find the under-utilised and the thrashing regions.

5.4   Networks with multiple job types

The restriction that all jobs are statistically identical can be relaxed considerably, in both open and closed networks. Here we shall assume that there are R different job types, and that the paths followed by type r jobs are governed by a separate routing matrix, Q_r $(r = 1, 2, \ldots, R)$. The (i, j)th element of that matrix, q_{rij}, gives the probability that a job of type r, having completed service at node i, goes to node j. The network can now be open with respect to all job types, closed with respect to all job types, or mixed, i.e. open for some types and closed for others (the definitions of 'open' and 'closed' are the same as before).

Having allowed the possibility of different routing patterns for the different job types, it is tempting to suppose that the latter may also have arbitrary required service time distributions, and that scheduling strategies other than FIFO may be in operation at certain nodes. Unfortunately, one cannot go too far in that direction and yet remain within the realm of solvable models. The following constraints have to be imposed on the nodes where queuing takes place:

1. If the scheduling strategy at node i is FIFO, then the required service times of all jobs (on each visit to that node) must be exponentially distributed, with the same mean, $1/\mu_i$.
2. If the required service times at node i have different distribution functions for the different job types – say $F_{ri}(x)$ for type r, with mean $1/\mu_{ri}$ – then the scheduling strategy there must belong to a special class of strategies. It must be 'symmetric'. This is a concept that requires elaboration.

To define the class of symmetric scheduling strategies, it is convenient to think of the queue as a sequence of numbered positions which jobs may occupy. When there are j jobs present, positions $1, 2, \ldots, j$ are occupied. An incoming job which finds that state enters position k ($k = 1, 2, \ldots, j + 1$) with probability $\alpha(k, j + 1)$. Unless $k = j + 1$, such an entry displaces the jobs in positions $k, k + 1, \ldots, j$ into positions $k + 1, k + 2, \ldots, j + 1$, respectively. The server may devote different fractions of its processing capacity to jobs in different positions. More precisely, with j positions occupied, a fraction $\beta(k, j)$ is devoted to position k ($k = 1, 2, \ldots, j$). If the job in position k leaves ($k < j$), the ones in positions $k + 1, k + 2, \ldots, j$ move to positions $k, k + 1, \ldots, j - 1$, respectively.

A symmetric scheduling strategy is one for which the probability of joining position k is equal to the fraction of processing capacity devoted to position k:

$$\alpha(k, j) = \beta(k, j); \quad k = 1, 2, \ldots, j; \ j = 1, 2, \ldots . \tag{44}$$

For instance, the processor-sharing strategy of section 4.4 is symmetric. It corresponds to the special case where a job may join any available position with equal probability, and equal fractions of processing capacity are allocated to all occupied positions: $\alpha(k, j) = \beta(k, j) = 1/j$. Another well-known symmetric strategy is 'last-come-first-served preemptive-resume', or LCFS PR. This strategy represents the operation of a stack. The server deals exclusively with the job in position 1, if any; a newly arriving job goes directly into position 1, interrupting the service of any job that might have been there. That is, $\alpha(1, j) = \beta(1, j) = 1$; $\alpha(k, j) = \beta(k, j) = 0$ ($k = 2, 3, \ldots, j$).

A characteristic of all symmetric scheduling strategies is that new arrivals start receiving service immediately, although they may not get the full attention of the server. The FIFO strategy is not symmetric (since new jobs join the tail, while service is given only to the head of the queue) and neither is any strategy which allocates service capacity on the basis of the types of the jobs present (e.g. a priority strategy).

Queueing networks with multiple job types, where all nodes are either delay nodes or single-server ones satisfying conditions 1 and 2, have product form solutions and are amenable to mean value analysis. The procedures for obtaining performance measures are similar to the ones described in sections 5.1 and 5.2, so our presentation here will be less detailed. All networks will be assumed to be in the steady-state.

Consider first the open case. Jobs of type r arrive externally into node i in a Poisson stream with rate γ_{ri} ($r = 1, 2, \ldots, R; i = 1, 2, \ldots, N$). For each r, at least one of the rates γ_{ri} is non-zero. The total rate at which type r jobs arrive into node i, λ_{ri}, can be determined by solving the following set of traffic equations.

$$\lambda_{ri} = \gamma_{ri} + \sum_{j=1}^{N} \lambda_{rj} q_{rji}; \qquad r = 1, 2, \ldots, R; i = 1, 2, \ldots, N. \quad (45)$$

In fact, what we have here is R independent sets of equations: one for each job type. Having determined λ_{ri} from (45), the load of type r at node i is equal to $\rho_{ri} = \lambda_{ri}/\mu_{ri}$ (of course, if node i is a single-server FIFO node then $\mu_{ri} = \mu_i$ for all r). The total load at node i is $\rho_i = \rho_{1i} + \rho_{2i} + \cdots + \rho_{Ri}$. In order that steady-state exists, the inequalities $\rho_i < 1$ must be satisfied for all single-server nodes.

Node i can be treated as an isolated and independent queueing system subjected to the loads ρ_{ri} ($r = 1, 2, \ldots, R$). If that is a delay node, i.e. if there are infinitely many servers available, then the average number of type r jobs in it, L_{ri}, and the average time type r jobs spend in it (per visit), W_{ri}, are equal to $L_{ri} = \rho_{ri}$ and $W_{ri} = 1/\mu_{ri}$, respectively. In the case of a single-server node, the corresponding expressions are

$$L_{ri} = \frac{\rho_{ri}}{1 - \rho_i}; \quad r = 1, 2, \ldots, R. \tag{46}$$

$$W_{ri} = \frac{1}{\mu_{ri}(1 - \rho_i)}; \quad r = 1, 2, \ldots, R. \tag{47}$$

These results are valid for the FIFO scheduling strategy, provided that all service times are exponentially distributed with the same mean. They are also valid for any symmetric strategy, with arbitrary distributions of the required service times. In the processor-sharing case, they were derived in 4.(52) and 4.(51).

The average number of visits that a job of type r makes to node i, v_{ri}, is given by an expression analogous to (9):

$$v_{ri} = \frac{\lambda_{ri}}{\gamma_r}; \quad r = 1, 2, \ldots, R; i = 1, 2, \ldots, N, \tag{48}$$

where $\gamma_r = \gamma_{r1} + \gamma_{r2} + \cdots + \gamma_{rN}$ is the total external arrival rate for type r (the type r throughput). The total average amount of service that a job of type r requires from node i is $D_{ri} = v_{ri}/\mu_{ri} = \rho_{ri}/\gamma_r$.

The total average time that a type r job spends at node i during its life in the network, B_{ri}, is given by

$$B_{ri} = v_{ri} W_{ri} = \begin{cases} D_{ri} & \text{if node } i \text{ is a delay node} \\ \dfrac{D_{ri}}{1 - \rho_i} & \text{if it is a single-server node} \end{cases} \qquad (49)$$

The total average response time for type r, W_r, is equal to

$$W_r = \sum_{i=1}^{N} B_{ri}; \quad r = 1, 2, \dots, R. \qquad (50)$$

Average sojourn times conditioned upon the initial node can be obtained as was done in the derivation of (8).

Example

Two computer sites are connected by a communication channel which allows jobs to be transferred from one to the other. There are three job types, with the following behaviour:

Type 1. External Poisson arrivals into site 1 with rate 5 jobs per minute. The average required service time there is 4 seconds. After completion of service at site 1, these jobs leave the network.

Type 2. External Poisson arrivals into site 2 with rate 10 jobs per minute. Average required service time there is 3 seconds. Departure from the network after completion of service at site 2.

Type 3. External Poisson arrivals into site 1 with rate 3 jobs per minute. Service at that site is followed by a transfer to site 2, with probability 0.6, and by a departure from the network with probability 0.4. After site 2, these jobs return to site 1. Their average required service times at sites 1 and 2 are 4 and 6 seconds, respectively.

The flow of jobs in this network is illustrated in figure 5.13.

Figure 5.13

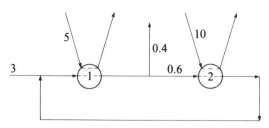

Assume that at both nodes there is a single server using a processor-sharing scheduling strategy. The performance measures of interest are the average response times for the three job types, W_1, W_2 and W_3.

Obviously, type 1 contributes only to the load at node 1. Similarly, type 2 contributes only to the load at node 2. Those contributions are $\rho_{11} = 5 \times (4/60) = 1/3$ and $\rho_{22} = 10 \times (3/60) = 0.5$, respectively. Jobs of type 3 circulate among the two nodes. The traffic equations for type 3 are

$$\lambda_{31} = 3 + \lambda_{32},$$

$$\lambda_{32} = 0.6\lambda_{31}.$$

These yield $\lambda_{31} = 7.5$ and $\lambda_{32} = 4.5$. Hence, the type 3 loads at the two nodes are $\rho_{31} = 7.5 \times (4/60) = 0.5$ and $\rho_{32} = 4.5 \times (6/60) = 0.45$, respectively. Steady-state exists because the total loads at nodes 1 and 2, $\rho_1 = 5/6$ and $\rho_2 = 0.95$, are both less than 1.

The average response times for types 1 and 2 are obtained immediately, since those jobs visit their respective nodes only once. Applying (47), we get

$$W_1 = \frac{\frac{4}{60}}{1 - \frac{5}{6}} = 0.4 \text{ minutes}; \quad W_2 = \frac{\frac{3}{60}}{1 - 0.95} = 1 \text{ minute}.$$

In order to determine W_3, we need the average number of visits that type 3 jobs make to nodes 1 and 2, v_{31} and v_{32}. These are given by (48):

$$v_{31} = \frac{7.5}{3} = 2.5; \quad v_{32} = \frac{4.5}{3} = 1.5.$$

Next, (49) yields the total average times that type 3 jobs spend at nodes 1 and 2:

$$B_{31} = \frac{2.5 \times \frac{4}{60}}{1 - \frac{5}{6}} = 1 \text{ minute}; \quad B_{32} = \frac{1.5 \times \frac{6}{60}}{1 - 0.95} = 3 \text{ minutes}.$$

Consequently, the average response time for jobs of type 3 is equal to $W_3 = 4$ minutes.

Suppose now that we are dealing with a closed network. All external arrival rates are 0 and all the row-sums of all routing matrices are 1. The job population in the network is fixed, and is composed of K_1 jobs of type 1, K_2 jobs of type 2, ..., K_R jobs of type R. Jobs leave the network and are immediately replaced by new ones, using entry points which may be different for the different types. All nodes are either delay nodes or single-server ones satisfying the conditions 1 and 2 which were discussed earlier.

The traffic equations for the closed network are

$$\lambda_{ri} = \sum_{j=1}^{N} \lambda_{rj} q_{rji}; \quad r = 1, 2, \ldots, R; \; i = 1, 2, \ldots, N. \tag{51}$$

Being homogeneous, these equations do not determine the arrival rates uniquely. They do so only up to R multiplicative constants: one for each job type. The average numbers of visits that jobs of type r make to the various nodes, v_{ri}, also satisfy equations (51). However, these last averages can be determined uniquely, because one of them is known, for each r. In particular, if the type r entry point is on the arc from node a to node b, then v_{ra} is given by

$$v_{ra} = \frac{1}{q_{rab}}$$

(the argument from section 5.2 applies without change).

Having obtained the average numbers of visits, v_{ri}, we can write recurrence mean value equations for the type r throughput, T_r, average number of jobs at node i, L $_{ri}$ and average time spent on each visit to node i, W_{ri}. These quantities depend on the population size vector, $K = (K_1, K_2, \ldots, K_R)$. The result that enables the recurrence equations to be written is a relation between the number of type s jobs seen at node i by an incoming type r job, $Y_{si}^{(r)}$ ($s = 1, 2, \ldots, R$), and the similar number seen by a random observer, L_{si}. That relation, which is analogous to (23), can be stated as follows:

$$Y_{si}^{(r)}(K) = L_{si}(K - u_r); \qquad s = 1, 2, \ldots, R; \; i = 1, 2, \ldots, N, \tag{52}$$

where u_r is the rth R-dimensional unit vector, i.e. the vector whose rth element is 1 and all other elements are 0. What (52) says is that a type r job coming into node i sees the state that a random observer would see if there was one less type r job in the network.

The recurrence equations resulting from (52) have almost the same form as (24), (25) and (26). For $r = 1, 2, \ldots, R$ we have

$$W_{ri}(K) = \frac{1}{\mu_{ri}} \left[1 + \delta_i \sum_{s=1}^{R} L_{si}(K - u_r) \right]; \qquad i = 1, 2, \ldots, N. \tag{53}$$

$$T_r(\mathbf{K}) = \frac{K_r}{\sum_{i=1}^{N} v_{ri} W_{ri}(\mathbf{K})} \, . \tag{54}$$

$$L_{ri}(\mathbf{K}) = T_r(\mathbf{K}) v_{ri} W_{ri}(\mathbf{K}); \quad i = 1, 2, \dots, N. \tag{55}$$

The definition of δ_i is the same as before. The fact that (53) holds for single-server nodes with different service requirements is a consequence of the properties of the symmetric scheduling strategies.

The initial conditions are $L_{ri}(0) = 0$ $(r = 1, 2, \dots, R; \; i = 1, 2, \dots, N)$. Note that for each population vector there are $(2N + 1)R$ equations. Also, in order to reach the desired population sizes, a total of $(K_1 + 1)(K_2 + 1) \cdots (K_R + 1)$ intermediate population vectors have to be examined. Clearly, when there are many job types and large populations for each type, the computational task is formidable.

As in the single type case, adequate approximations can be obtained with much less effort by turning the recurrence equations into fixed-point ones. This is achieved by relating the averages associated with the population vector $\mathbf{K} - \mathbf{u}_r$ to those associated with the vector $\mathbf{K}$. One such relation which gives reasonably accurate results is the following:

$$L_{si}(\mathbf{K} - \mathbf{u}_r) = \begin{cases} L_{si}(\mathbf{K}) & \text{if } s \neq r \\ \dfrac{K_r - 1}{K_r} L_{ri}(\mathbf{K}) & \text{if } s = r \end{cases} ;$$

$$s = 1, 2, \dots, R; \; i = 1, 2, \dots, N. \tag{56}$$

The total average time that a type r job spends at node i, $B_{ri}(\mathbf{K})$ and the total average time that it spends in the network, $W_r(\mathbf{K})$, are obtained from

$$B_{ri}(\mathbf{K}) = v_{ri} W_{ri}(\mathbf{K}) \tag{57}$$

and

$$W_r(\mathbf{K}) = \sum_{i=1}^{N} B_{ri}(\mathbf{K}) = \frac{K_r}{T_r(\mathbf{K})} \, . \tag{58}$$

If, instead of being given the routing matrices as network parameters, one has the total average service requirements for type r at node i, D_{ri}, then equations (53), (54) and (55) become slightly simpler:

$$B_{ri}(\mathbf{K}) = D_{ri}\left[1 + \delta_i \sum_{s=1}^{R} L_{si}(\mathbf{K} - \mathbf{u}_r) \right]; \quad i = 1, 2, \dots, N. \tag{59}$$

$$T_r(\mathbf{K}) = \frac{K_r}{\sum_{i=1}^{N} B_{ri}(\mathbf{K})} \, . \tag{60}$$

$$L_{ri}(\mathbf{K}) = T_r(\mathbf{K}) B_{ri}(\mathbf{K}); \quad i = 1, 2, \dots, N. \tag{61}$$

If the network is mixed, then the recurrence is with respect to the populations of the closed job types only. The state of an open job type, s, seen at node *i* by an incoming job of type *r* (regardless of whether the latter belongs to an open or closed type), is the same as the corresponding state seen by a random observer:

$$Y_{si}^{(r)}(\boldsymbol{K}) = L_{si}(\boldsymbol{K}) \text{ when } s \text{ is open}; \quad i = 1, 2, \ldots, N. \tag{62}$$

(The vector $\boldsymbol{K}$ describes the population sizes of the closed types only.)

Finally, let us mention two further generalisations that are possible in queueing network models, but have not been covered here. First, the service at a node may be provided by a server whose speed depends on the number of jobs present. The nodes with a single server of constant speed, those with a finite number of parallel servers and those with infinitely many parallel servers can be considered as special cases of a node with a state-dependent server.

Second, the type of a job may change as it follows its path through the network. By using such changes, and by introducing artificial job types, one could extend considerably the way the routing of jobs is described. For instance, one could have deterministic routes through the network, or routes whose future depends on their past.

The price of adopting either of these generalisations is that the mean value analysis no longer applies. The product form of the joint network state distribution has to be used, with all the attending difficulties of computing normalisation constants (in the case of closed networks).

Exercises

1. Turn the example in this section into a closed network by replacing the external arrivals and departures with feedback loops. Solve the resulting model for different population vectors (K_1, K_2, K_3).

2. Construct a closed network model of a multiprogrammed computer (along the lines of the network in figure 5.7), where the jobs may be of different types. For instance, there could be 'CPU-bound jobs' and 'I/O-bound jobs'. Choose a configuration of manageable size, select values for the parameters and determine the corresponding performance measures.

Literature

Jackson's original paper [14] was motivated mainly by problems in manufacturing, where components go through several stages of operation in the course of a production line. Most of the major advances in the theory, such as multiple job types and different scheduling strategies and required service times distributions, have occurred relatively recently. The computing field has provided much of the impetus for new developments. There is a number of books covering queueing network material. Lavenberg [23] and Lazowska *et al.* [24] are rather practical in orientation, while Sauer and Chandy [30] and Kelly [16] deal with various aspects of the theory. The numerical implementation of product form solutions for closed networks is considered in Bruell and Balbo [1]. The decomposition approximation method is treated in Courtois [6]. The control of multiprogramming is discussed in Gelenbe and Mitrani [11].

6
Packet-switching networks

A communication system can be regarded as a collection of channels, or links, by means of which messages can be passed around a network of source and destination nodes. The transmission and acknowledgement of messages is governed by a set of rules which is usually referred to as the 'network protocol'.

One way of satisfying a communication request is to establish a path, or 'circuit' consisting of one or more links between the source and the destination, and dedicate it exclusively to that request for its entire duration. Networks built on that principle are called 'circuit-switching'. Alternatively, links may be shared by several communication tasks proceeding in parallel. That usually involves dividing long messages into smaller units, or 'packets', and interleaving the transmissions of packets from different messages along the same link. Such networks are called 'packet-switching'.

Packet-switching is generally a more cost-effective method of providing communication services than circuit-switching. Most of the data traffic in the world is currently carried by packet-switching networks, and there is a tendency to packetise voice messages, too. In this chapter, we shall examine the performance of several packet-switching networks. Our aim will be to illustrate the modelling techniques that are used, rather than to cover the topic exhaustively (the latter objective would require much more than one chapter). In all cases, the networks will be 'single-hop', i.e. such that every source-destination pair is connected by one channel. Within that class, there are two broad categories of networks: those with distributed control, where each source makes its own decisions about when to occupy the channel, and those where the channel use is controlled centrally. We shall start with the first category.

6.1   Broadcast networks

The simplest way of controlling a channel is not to control it at all. The moment a packet is ready for transmission at any node, let it be transmitted. That policy was adopted in the famous ALOHA network, developed at the University of Hawaii in the early 1970s. There the communication channel is a given radio frequency, and to transmit a packet is to broadcast it. All nodes listen to the channel all the time; a destination recognises the packets intended for it by an appropriate address field.

The price paid for the simplicity of this channel sharing method is that not all transmissions are successful. If the broadcasts of two or more packets overlap, the information contained in those packets is corrupted. Such an event is called a 'collision'. When a collision occurs, the senders that are involved in it soon become aware of the fact (because they, too, are listening to the channel). All participants then back off and attempt to retransmit their packets after random periods of time. That randomness is essential: without it, a collision would necessarily be followed by another collision.

The first performance question that arises in connection with the ALOHA protocol concerns the expected maximum achievable throughput of the channel. Assume, for simplicity, that all packets are of fixed length and choose the time unit so that the packet transmission time is 1. Then, if there were no collisions and packets were transmitted one after the other without any gaps, the throughput would be 1 packet per unit time. How close to that ideal can one expect to get in reality?

The stream of arrivals at the channel consists of newly generated packets and of old ones, offered for retransmission because of past collisions. Denote the total arrival rate by G (packets per unit time). Let S be the average number of successfully transmitted packets per unit time, i.e. the throughput. Then, if q is the probability that an incoming packet is transmitted successfully, we can write

$$S = Gq. \tag{1}$$

In order to estimate the probability q, assume that the total arrival stream is Poisson, with rate G. This is not necessarily correct, but will do as a first approximation. Now, a packet arriving at time t will be successful if its transmission does not overlap with that of any other packet, i.e. if no other packets arrive in the interval $(t - 1, t + 1)$. This situation is illustrated in figure 6.1.

Figure 6.1

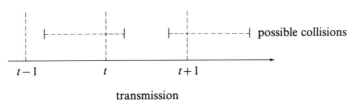

transmission

From the properties of the Poisson process (section 2.2) it follows that the probability of no arrivals during the interval $(t-1, t+1)$ is independent of t and is equal to $q = e^{-2G}$. Hence, the relation (1) between the offered traffic rate and the throughput becomes

$$S = Ge^{-2G}. \tag{2}$$

That relation is plotted in figure 6.2.

Figure 6.2

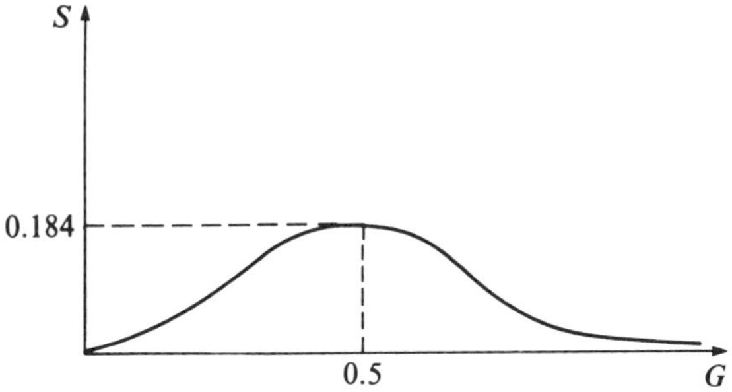

S

0.184

0.5 G

To find the maximum value that S can attain, take the derivative of the right-hand side of (2) with respect to G and equate it to 0. This immediately reveals that the optimal traffic rate is $G = 1/2$, and that the corresponding maximal throughput is $S = 1/(2e)$.

We have arrived at the somewhat disappointing conclusion that, with the free-for-all ALOHA protocol, one cannot expect to achieve more than about 18 % of the ideal one-packet-per-unit-time throughput. Moreover, any attempt to get more packets through by increasing the offered traffic rate is bound to misfire eventually and result in a dramatic drop in throughput. This situation is reminiscent of the thrashing phenomenon in multiprogrammed computer systems (section 5.3).

The poor utilisation of the available channel capacity is due to packet collisions (and the consequent necessity for retransmission). It is natural, therefore, to attempt to improve matters by reducing the period of time during which collisions are possible. One way of achieving this is to allow transmissions to start only at selected points in time, say the integer points $\{0, 1, 2, \ldots\}$. Thus, if a packet arrives sometime during the interval $(n - 1, n)$, it will start its transmission at time n and will complete it at $n + 1$. That transmission will be successful if no other packets are transmitted at the same time.

This modified protocol is referred to as 'slotted ALOHA'; the intervals $(n, n + 1)$ $(n = 0, 1, \ldots)$ are called 'slots'. It is readily seen that the effect of slotting is to double the maximum achievable throughput. Indeed, a packet arriving in a given slot will avoid collision if, and only if, there are no other arrivals in the same slot. The probability of that event is $q = e^{-G}$. Thus, in the slotted ALOHA system, the throughput is related to the offered traffic rate as follows:

$$S = Ge^{-G}. \tag{3}$$

The function (3) is similar in shape to the one plotted in figure 6.2. However, it reaches a maximum of $1/e$ (approximately 0.36), at $G = 1$. This is a two-fold increase, compared with the original ALOHA system.

A further improvement in performance can be achieved by making a better use of the fact that all nodes are able to monitor the channel continuously. Suppose that a transmission can be started at any time but only if the sender, having listened to the channel, has found it idle. If another transmission is sensed to be in progress, the sender backs off and tries again after a random period of time. Such a protocol is commonly known as 'carrier sense multiple access', or CSMA. Perhaps the best-known example of a CSMA network is Ethernet, where the channel is a cable connecting all nodes.

Collisions may still occur under a CSMA protocol. Because of the finite speed at which signals are propagated, a sender may think that the channel is idle, whereas in fact another transmission has just started. Let d be the propagation delay of the channel, i.e. the time for the signal to travel from one station to another. If a transmission is attempted at time t, and another one starts at some point in the interval $(t - d, t + d)$, then the later of the two senders will not be aware of the earlier one, and the result will be a collision. Consequently, a transmission started in the belief that the channel is idle will be successful if, and only if, there are no other arrivals in an interval of length $2d$. That is, $q = e^{-2dG}$.

To get a relation between the throughput, S, and the offered traffic rate, G, we argue as follows. Since the transmission time is 1, S is equal to the load (work brought in per unit time) due to the successful packets. Hence, it is also equal to the probability, or the fraction of time, that the channel is busy with a successful transmission. According to the CSMA protocol, no one attempts to transmit during the successful busy time. During the rest of the time, such attempts are made at rate G. Therefore, the overall rate at which transmissions are started is $(1 - S)G$. Remembering that each attempt is successful with probability e^{-2dG}, we can write

$$S = (1 - S)Ge^{-2dG}.$$

Solving this equation for S yields

$$S = \frac{Ge^{-2dG}}{1 + Ge^{-2dG}}. \tag{4}$$

The function in the right-hand side of (4) reaches a maximum of $1/(1 + 2de)$, for $G = 1/(2d)$. This is a much better state of affairs than in the other two systems, since the value of d is usually quite small. Of course, if the offered traffic rate G is allowed to exceed the optimal value by much, the throughput will again drop to something near zero.

This brings us to the question of existence of steady-state in a broadcast network. Given the rate at which new packets are generated by the user population, what happens to the total offered traffic rate in the long run? Does it remain bounded and reasonably low, so that an acceptable throughput is achieved, or does it keep growing, eventually bringing the throughput down to zero?

To get some idea of the phenomena involved, consider the slotted ALOHA system. Assume that new packets (whose transmission has not been attempted before) arrive in a Poisson stream with rate λ. Packets that have suffered at least one collision and have not yet been transmitted are said to be 'blocked'. While the packet remains blocked, it makes an attempt to transmit in each subsequent slot with a fixed probability, α.

The state of the system at the beginning of the nth slot is described by a single integer, X_n, which represents the number of blocked packets at that moment. The above assumptions imply that the process $X = \{X_n; n = 0, 1, \ldots\}$ is a Markov chain whose possible states are the non-negative integers. The behaviour of that chain is characterised by its one-step transition probabilities,

$$q_{ij} = P(X_{n+1} = j \mid X_n = i). \tag{5}$$

Suppose that the system is in state i at time n, i.e. there are i blocked packets. The state at time $n + 1$ will depend on what happens during the interval $(n, n + 1)$. There are several possibilities:

(a) If k new packets arrive during the nth slot, where $k \geqslant 2$, then they will collide and the chain will jump to state $i + k$.

(b) If 1 new packet arrives and at least one of the blocked packets attempts to transmit, then again there will be a collision and the next state will be $i + 1$.

(c) If either 1 new packet arrives and no blocked packets attempt to transmit (in which case the new arrival will be successful), or no new packets arrive and any number other than 1 of the blocked packets attempt to transmit, then the state will remain i.

(d) If no new packets arrive and exactly one of the blocked packets attempts to transmit, then the attempt will be successful and the next state will be $i - 1$.

Thus, the transition probabilities q_{ij} can be expressed in terms of the probability, α, with which blocked packets attempt to transmit, and the Poisson arrival probabilities:

$$a_k = P(k \text{ arrivals during a slot}) = \frac{\lambda^k}{k!} e^{-\lambda}; \quad k = 0, 1, \ldots . \tag{6}$$

It is easy to verify that the expressions corresponding to the four event categories (a)–(d) are

$$q_{i,i+k} = \begin{cases} a_k & \text{for } k = 2, 3, \ldots , \\ a_1[1 - (1 - \alpha)^i] & \text{for } k = 1, \\ a_1(1 - \alpha)^i + a_0[1 - i\alpha(1 - \alpha)^{i-1}] & \text{for } k = 0, \\ a_0 i\alpha(1 - \alpha)^{i-1} & \text{for } k = -1. \end{cases} \tag{7}$$

When the one-step transition is from state i to state j, we say that the chain makes a jump of size $j - i$. Clearly, the possible jump sizes are $-1, 0, 1, 2, \ldots$. The average jump size when the chain is in state i is called the 'drift at state i' and is denoted by D_i. For any given i, the drift is computed according to the formula for the mean,

$$D_i = \sum_{k=-1}^{\infty} k q_{i,i+k}. \tag{8}$$

A numerical evaluation of D_i is easily carried out, using the expressions (7). A typical plot of the drift at state i, regarded as a function of i, is shown in figure 6.3.

Figure 6.3

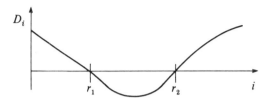

Provided the retransmission probability α is not too large, the drift function has two roots, which we have denoted in the figure by r_1 and r_2. These two roots divide the state space of the Markov chain into three regions with different properties:

(i) When the system state, i, is less than r_1, the drift is positive. That means that the number of blocked packets has a tendency to increase, pushing the state in the direction of r_1.

(ii) When i is between r_1 and r_2, the drift is negative. The number of blocked packets tends to decrease, pulling the state back towards r_1.

(iii) When i is greater than r_2, the drift is again positive. The number of blocked packets tends to grow, pushing the state further away from r_2. Moreover, the more the system drifts to the right, the more certain it is to continue to do so. This brings about an explosion of blocked packets and a collapse of throughput (since every slot results in a collision).

Thus, regions (i) and (ii) are stable, with the system state hovering around point r_1. Region (iii) is unstable. Once in it, the state quickly drifts away to infinity.

The above observations allow us to conclude that steady-state does not exist for this Markov chain. Indeed, since every state is reachable from every other state, one can claim that no matter what the initial state, and no matter how attractive the pull of point r_1, the chain will find itself eventually in region (iii). It will then drift towards saturation. Hence, the long-run probability of any finite state is zero.

Having reached this conclusion, we might wash our hands and dismiss the slotted ALOHA protocol from further consideration. To do that, however, would be to ignore the practical aspects of the protocol's performance. The fact is that the process can remain in the stable regions for a very, very long time before moving into the unstable one. The combinations of circumstances leading to instability are so rare that special techniques have to be employed to estimate the length of the period until their occurrence. They do not show up in simulations. A recent study has shown, for example, that if the new packet arrival rate is $\lambda = 0.1$ and the retransmission probability is $\alpha = 0.001$, and if the system starts at a point in the stable region, then the average period until the onset of instability is on the order of e^{246} slots. Such a system is, to all practical purposes, stable.

The behaviour described above is common to all broadcast networks (including CSMA ones), where blocked packets make retransmission attempts independently of each other. The reason for the instability is of course that, when the number of blocked packets increases beyond a certain level, there are so many collisions that the backlog cannot be cleared. As a way out of this, it has been suggested that the average interval between successive retransmission attempts by a blocked packet should increase with the number of collisions the packet has suffered. For instance, every time a blocked packet makes an unsuccessful attempt to transmit, it doubles the average period until the next attempt. This last policy is referred to as 'exponential back-off'. The rationale behind it is that, even though the number of blocked packets may temporarily become very large, the total offered traffic rate will not grow unduly, the probability of collision will not be large and the throughput will be sufficient to clear the backlog.

For a long time it was thought that under the exponential back-off policy, and with suitable parameter values, a broadcast network reaches steady-state. Both observations and simulations confirmed this. However, in 1985 it was demonstrated that even in this case the unstable region exists and is eventually entered. Admittedly, the period to instability is so large that the fact that it is finite is of theoretical interest only.

In order to eliminate the unstable region completely, a much tighter control on retransmissions has to be exercised. The net effect of the control policy should be that the average retransmission intervals of a blocked packet increase in proportion to the number of blocked packets. Then the offered traffic rate from the set of blocked packets will remain bounded, regardless of the size of that set. It is possible to ensure that the channel capacity available to new packets does not drop below a certain minimum and, if the arrival rate of new packets is less than that minimum, steady-state will exist.

In some situations it is more realistic to assume a finite source of new packets, rather than the infinite one implied by the Poisson stream. As an example of such a system, consider a CSMA network, such as Ethernet, where new packets are generated by K identical terminals. Each terminal, having submitted a packet, waits until it is successfully transmitted, then thinks for a while, before submitting the next packet.

This system can be modelled by a closed queueing network with three nodes and K jobs (packets), forever circulating among them. Node 1 represents the think state; jobs remain there for a random think time which we shall assume is distributed exponentially with mean $1/\gamma$. Node 2 is the communication channel, a single server. At most one job can be at node 2 at any one time. Service times are distributed exponentially with mean $1/\mu$ (this departure from the assumption of fixed packet lengths is made in order to simplify the analysis). Node 3 represents the blocked state. Jobs enter there either after finding the channel busy or after a collision. In both cases, jobs remain at node 3 for a random period distributed exponentially with mean $1/\xi$. The destination of a job leaving either node 1 or node 3 depends on the state of node 2: if the channel is free, the job attempts to occupy it; otherwise the job joins node 3.

Every job attempting to join node 2, whether it does so from node 1 or from node 3, has a fixed probability β, of being involved in a collision. We shall see later how that probability is determined. If a collision occurs, the attempt is unsuccessful and the job joins node 3. After a service completion at node 2, a job returns to node 1. The flow of jobs among the three nodes is illustrated in figure 6.4.

Figure 6.4

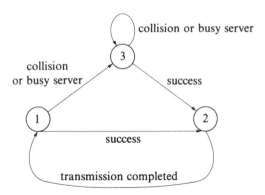

The state of the system at any moment in time is described by specifying the number of jobs, i, currently at node 2 (that number can be either 0 or 1), and the number of jobs, n, currently at node 3 ($n = 0, 1, \ldots, K - i$). The number of jobs at node 1 is then equal to $K - n - i$. Under the assumptions that we have made, this state process is Markov. Since the number of possible states is finite, steady-state always exists. Let $p_i(n)$ be the steady-state probability that there are i jobs at node 2 and n jobs at node 3 ($i = 0, 1$; $n = 0, 1, \ldots, K - i$). These probabilities can be found by writing and solving a set of balance equations.

Suppose that the system is in state $(0, n)$, i.e. the channel is idle and there are n jobs at node 3. To get out of that state, either one of the $K - n$ jobs at node 1 has to stop thinking, or one of the n jobs at node 3 has to make an attempt to join node 2 and be successful. Hence, the instantaneous transition rate out of state $(0, n)$ is $(K - n)\gamma + n\xi(1 - \beta)$. Transitions into state $(0, n)$ can take place from state $(1, n)$, if the service in progress completes (rate μ), or from state $(0, n - 1)$, if one of the $(K - n + 1)$ jobs at node 1 makes an unsuccessful attempt to join node 2 (rate $(K - n + 1)\gamma\beta$). Thus we have the balance equations

$$[(K - n)\gamma + n\xi(1 - \beta)]p_{0n} = \mu p_{1n} + (K - n + 1)\gamma\beta p_{0,n-1};$$
$$n = 0, 1, \ldots, K. \quad (9)$$

(When $n = 0$, the term involving $p_{0,-1}$ is 0 by definition. Also, $p_{1K} = 0$ by definition.)

Similarly, by balancing the transition flows out of, and into, state $(1, n)$, we obtain the equations

$$[(K - n - 1)\gamma + \mu]p_{1n} = (K - n)\gamma p_{1,n-1} + (K - n)\gamma(1 - \beta)p_{0n}$$
$$+ (n + 1)\xi(1 - \beta)p_{0,n+1};$$
$$n = 0, 1, \ldots, K - 1 \quad (10)$$

(by definition, $p_{1,-1} = 0$).

Using equations (9) and (10) alternately, for $n = 0, 1, \ldots, K$, one can express all the unknown probabilities in terms of p_{00}. That last probability is determined by the normalising equation

$$\sum_{n=0}^{K} p_{0n} + \sum_{n=0}^{K-1} p_{1n} = 1. \quad (11)$$

Having found the steady-state probabilities p_{in}, we obtain the channel throughput, S, as the average number of service completions at node 2 per unit time:

$$S = \mu \sum_{n=0}^{K-1} p_{1n}. \quad (12)$$

The total offered traffic rate, G, is equal to the average number of attempts made to join node 2 per unit time:

$$G = \sum_{n=0}^{K} [(K - n)\gamma + n\xi]p_{0n}. \quad (13)$$

The average response time for a packet, W, i.e. the interval between leaving the think state and completing service at node 2, is obtained by applying Little's result to node 1, and also to the sub-system consisting of nodes 2 and 3, and then summing the resulting equations (an almost identical derivation led to 5.(43)). This yields

$$K = S\left(W + \frac{1}{\gamma}\right),$$

or

$$W = \frac{K}{S} - \frac{1}{\gamma} \tag{14}$$

(see also the derivation of 3.(47) and 3.(48)).

Let us return now to the problem concerning the collision probability, β. We have analysed the model assuming that the value of β is known, but in fact that value is one of the unknown quantities which need to be determined. A more basic parameter of the network is the channel propagation delay, d. The value of d is given. An attempt made at time t to occupy the channel fails, resulting in a collision, if there is at least one other such attempt made in the interval $(t - d, t + d)$. If the total offered traffic rate due to the other $K - 1$ jobs is H, and if they make their attempts in a Poisson stream, then the probability of a collision with our attempt is $1 - e^{-2dH}$.

The above remarks suggest a simple approximate procedure for determining β. Remember that all performance measures that we have derived, including the total offered traffic rate, G, are functions of K and β (also of γ and μ, but those are regarded as fixed throughout). To make that dependence explicit, denote the right-hand side of (13) by $G_K(\beta)$. Now, since a job attempting to occupy the channel does not contribute to the colliding traffic rate, it is natural to estimate the latter by $G_{K-1}(\beta)$. That is, use as H the total offered traffic rate in a network with $K - 1$ jobs. This yields a fixed-point equation for β:

$$\beta = 1 - e^{-2dG_{k-1}(\beta)}. \tag{15}$$

In practice, (15) would be solved by iterations: start with an initial value for β, say 0; solve the model with $K - 1$ jobs and use (15) to get the next value of β; go on like this until the iterations converge to the fixed point. Finally, solve the model with K jobs.

The results obtained in this way are generally quite accurate, especially when the propagation delay is small. In particular, if $d = 0$, then the collision probability is also zero and equations (9), (10) and (11) yield the exact solution.

Exercises

1. Plot the slotted ALOHA throughput, S, given by equation (3), against the total offered traffic rate, G. Find the range of G values for which the throughput is at least 0.1 packets per unit time.

2. Consider a modification of the CSMA protocol whereby a packet arriving when the channel is busy attempts to transmit immediately after the completion of the current transmission. This version of the protocol is called 'persistent CSMA'. Of course, if two or more packets arrive during a transmission, they will all make their attempts as soon as it completes, which will cause them to collide.

Using an argument similar to the one leading to (4), derive a relation between throughput and offered traffic rate for the persistent CSMA. (Hint: an arrival during a successful transmission will itself be successful at the end of it if no other arrivals occur during that transmission; the probability of that event is e^{-G}).

3. Implement the iterative procedure for solving the finite source CSMA network. Evaluate the average response time for different values of the parameter ξ and observe the nature of the dependency.

6.2   Slotted rings

We shall now turn our attention to networks where the control of accesses to the channel is centralised, rather than distributed. The *raison d'être* of a central controller is to ensure that the channel capacity is shared in a fair and orderly fashion among the users. Collisions of packets are eliminated, a guaranteed throughput is maintained and the instability phenomenon is avoided. On the other hand, the control mechanism itself, and the overheads associated with it, take up some of the available channel capacity.

In all our examples, the communication channel is a closed wire loop (or an optical fibre one), which is referred to as the 'ring'. Signals travel along the ring in one direction only. A certain number of stations which may act as both sources and destinations of packets are connected to the ring. This general arrangement is illustrated in figure 6.5.

Figure 6.5

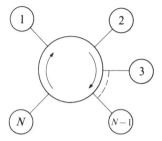

The present section is concerned with two slotted ring networks. Packets of information are carried in one or more frames, or 'slots', of fixed size. These slots circulate around the ring continuously. Senders write into them and receivers read from them. One of the stations on the ring is the central controller: it generates the empty slots and also carries out certain error detection and recovery functions.

A well-established slotted ring network is the 'Cambridge ring'. It is characterised by a set of hardware-enforced rules which are designed to prevent any user, or group of users, from either monopolising the channel or being starved of service. At any moment in time, a slot is occupied or free, depending on whether it is currently engaged in a transmission or not. A station which has packets to transmit examines every passing slot until a free one comes along. That slot is then filled with a packet, together with the addresses of the sender and receiver, and is marked 'occupied'. As it continues on its way around the ring, the slot is examined by each of the other stations in turn. The receiver, having recognised its address, copies the packet from the slot into its local buffer and indicates a successful reception by means of special response bits. The slot remains occupied until it returns back to the sender, who must mark it 'free'. It is this passing on of a free slot which ensures that every station gets its share of the channel capacity.

These rules imply that the transmission of a long message consisting of many packets is done in quanta, one packet at a time. After each quantum of service, the server is released. This is reminiscent of the round-robin scheduling strategy, which in turn approaches the processor-sharing one when the quantum size is small (see section 4.4). If several stations wish to transmit messages simultaneously, and if the respective destinations are willing to accept those messages, then all transmissions proceed in parallel and at equal rates.

The parameters characterising the channel are its bandwidth, b (bits passing through per unit time) and delay, τ (time for one bit to go all the way around the ring). The number of bits that the ring can carry at any moment is $B = b\tau$. Thus, if a slot contains w bits, the number of slots that can be accommodated is $S = \lfloor B/w \rfloor$, where $\lfloor x \rfloor$ is the integer part of x.

Consider a ring with S slots, and suppose that n stations wish to transmit simultaneously, each having an infinite supply of packets. Denote by $T_S(n)$ the corresponding channel throughput, i.e. the average number of packets that are delivered per unit time. The exact determination of that quantity is a rather complex combinatorial problem which is still open. However, when n is considerably larger than S, a good approximation can be obtained by a very simple argument. Let us follow the activities of one slot. After it is loaded with a packet, it makes a complete cycle around the ring, is released, travels to the next active station (average distance $1/n$th of the circumference) and is loaded with a packet again. Thus the average interval between consecutive packet loads is $(1 + 1/n)\tau = [(n + 1)/n]\tau$. Hence, the average number of packets delivered by one slot per unit time is $n/[(n + 1)\tau]$. Since there are S slots circulating around the ring, we have

$$T_S(n) = \frac{nS}{(n + 1)\tau}; \quad n > S. \tag{16}$$

It is easily seen that if $S = 1$, then this expression is exact for every n. A similar reasoning shows that when n is small and S is large, the ring throughput is approximated by

$$T_S(n) = \frac{nS}{(S + 1)\tau}; \quad n < S. \tag{17}$$

If $n = 1$ then (17) is exact for every S.

We conclude that an S-slotted ring can be modelled by a state-dependent processor-shared server which, when working on n jobs, has a service capacity given by (16) or (17). Suppose that K identical stations are connected to the ring, each with average think times $1/\gamma$ and average message lengths $1/\mu$. The situation is similar to the one illustrated in figure 5.11. Let p_i be the steady-state probability that i messages are being transmitted in parallel $(i = 1, 2, \ldots, K)$. These probabilities satisfy the following set of balance equations:

$$(K - i)\gamma p_i = T_S(i + 1)\mu p_{i+1}; \quad i = 0, 1, \ldots, K - 1. \tag{18}$$

The state diagram giving rise to the equations is shown in figure 6.6.

Figure 6.6

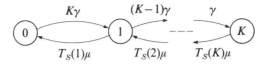

For the state process to be Markov, the think times and message lengths have to be exponentially distributed. However, the validity of equations (18) does not rely on such assumptions.

Having obtained the probabilities p_i from (18) and the normalising equation, the message throughput, T (the average number of message transmissions completed per unit time) is given by

$$T = \sum_{i=1}^{K} p_i T_S(i)\mu. \tag{19}$$

To find the average response time, W, we use the familiar expression (see (14))

$$W = \frac{K}{T} - \frac{1}{\gamma}. \tag{20}$$

On top of the hardware protocol that we have just described, the Cambridge ring has several higher level protocols. One of these, called the 'basic block' protocol, allows a sender to group a number of packets into a single block. Once a receiver has agreed to accept such a group, it refuses packets from other sources until the entire block has been received. Under this protocol, the ring may be shared between two or more transmissions only if they are to different destinations. For example, A may transmit a block to B in parallel with C transmitting one to D but, if A and B wish to send blocks to C, they have to do so in sequence.

The basic block protocol can be modelled by a closed queueing network with $K + 1$ nodes such as the one shown in figure 6.7.

Figure 6.7

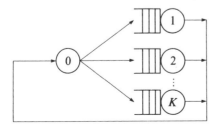

There are K jobs, or blocks, circulating among the nodes at all times. Node 0 represents the think state. A job leaving node 0 and joining node j ($j = 1, 2, \ldots, K$) corresponds to a station wishing to send a block to station j. There is a single server and a FIFO queue at each of the nodes $1, 2, \ldots, K$. Thus, if several blocks have been submitted for transmission to station j, one of them is being transmitted and the others are waiting in order of arrival. After completing service at node j, a job returns to node 0.

Note that, whereas a receiving station is represented by a node in our model, a sending one is associated with a job. To make that association clearer, the K jobs are numbered $1, 2, \ldots, K$. The average time that job i spends in the think state is $1/\gamma_i$ and the probability that after leaving node 0 it will join node j is q_{ij} ($i, j = 1, 2, \ldots, K$). If station i never sends messages to itself via the ring, then $q_{ii} = 0$. The lengths of the blocks destined for station j, i.e. of the jobs at node j, are distributed exponentially with mean $1/\mu_j$.

We seem to have here a standard closed queuing network with multiple job types, of the sort described and solved in section 5.4. However, the present model has an important distinguishing feature which precludes the direct application of standard solution techniques. Since the blocks at the heads of the queues $1, 2, \ldots, K$ (if any) are transmitted in parallel on the ring, the rate at which each transmission progresses depends on the number of non-empty queues (in fact, the number of waiting jobs also influences the service capacity, but we shall ignore that effect).

Suppose that there are n non-empty queues, that is n blocks being transmitted in parallel. To avoid having to consider separately the case where n is less than the number of slots, let us assume that the ring has a single slot. Then the packet throughput is given by (16), for $S = 1$. Therefore, the rate, $r(n)$, at which each of the n transmissions progresses is equal to

$$r(n) = \frac{T_1(n)}{n} = \frac{1}{(n + 1)\tau}. \tag{21}$$

If queue j is one of the non-empty ones, then the instantaneous completion rate for the job at its head is $r(n)\mu_j$.

This dependency of the service rate on the number of non-empty queues invalidates the known closed queuing network results. Fortunately, an approximate solution can be obtained quite easily by the fixed-point method.

The approximation consists of replacing the state-dependent service rates, $r(n)$, by a fixed one, r. The latter is then chosen so as to be consistent with the average performance of the ring.

If we assume that all transmissions progress at the rate of r packets per unit time, then the service time distribution at node j is exponential with mean $1/(r\mu_j)$, regardless of the state of the various queues. The network can now be solved, for instance by applying a mean-value iterative procedure (sections 5.2 and 5.4). That solution yields, among other things, the utilisation, U_j, of node j, for $j = 1, 2, \ldots, K$ (see expression 5.(33) and its obvious generalisation to multiple job types). Since U_j is the probability that queue j is non-empty, the average number of non-empty queues, $\bar{n}$, is equal to

$$\bar{n} = \sum_{j=1}^{K} U_j. \tag{22}$$

This value of $\bar{n}$ is of course a function of $r: \bar{n} = \bar{n}(r)$. A fixed-point equation for r is obtained by replacing, in (21), n by $\bar{n}$ and equating the result to r:

$$r = \frac{1}{[\bar{n}(r) + 1]\tau}. \tag{23}$$

An iterative solution procedure would start with an initial value for r, say $r_0 = 1/[(K + 1)\tau]$. Having solved the network, (23) is used to get the next value of r, and so on. In practice, reasonable approximations are obtained after a few iterations. Performance measures such as average response times are taken from the last network iteration.

Slot release at destination

One of the essential requirements of the Cambridge ring protocol was that an occupied slot should return to the sender before it is passed on free. By means of the returning slot, the receiver acknowledges or denies that the packet was received. However, if one is willing to forgo these acknowledgements, a more efficient use of the channel can be achieved by freeing a slot as soon as it has delivered its packet – at the destination node. In this way, one slot may be employed several times during a single trip around the ring. That is the philosophy adopted in the design of British Telecom's 'Orwell ring'.

Let us estimate the state-dependent packet throughput of a destination-release ring with S slots. Consider a configuration where there are n equidistant active stations, each having an infinite supply of packets to send. Assume that, for each sender, the destinations of consecutive packets are independent of each other and are uniformly distributed around the ring (i.e. there are many more possible destinations than active stations).

After being loaded with a packet, a slot travels to its destination where it is released and is then occupied again by the next active station along the ring. If the destination is between the $(i-1)$st and the ith active stations after the sender $(i = 1, 2, \ldots, n)$, then the distance travelled between two consecutive packet loads is a fraction i/n of the circumference of the ring. Since this occurs with probability $1/n$, the average interval of time between two consecutive packet loads for one slot is equal to

$$\sum_{i=1}^{n} \frac{i}{n^2} \tau = \frac{n+1}{2n} \tau.$$

Hence, the average number of packets delivered by one slot per unit time is $2n/[(n+1)\tau]$. Multiplying this by the number of slots yields the packet throughput, $T_S(n)$:

$$T_S(n) = \frac{2nS}{(n+1)\tau}. \tag{24}$$

Comparing this expression with the similar results for the Cambridge ring, (16) and (17), we see that the destination release of slots improves the throughput by a factor of two or more, at least when the destinations are uniformly distributed.

It should be pointed out, however, that it is now possible to prevent an unlucky station from occupying any slots and thus deprive it of service. An example of what may happen is illustrated in figure 6.8.

Figure 6.8

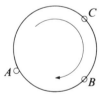

Station A fills empty slots with packets intended for B. Station B releases the slots, they return to A empty and are filled again. In consequence, station C never sees an empty slot.

One has to guard against such unfortunate situations. In the Orwell ring, the problem is avoided by forcing stations which 'hog' the channel to relinquish it from time to time.

Exercises

1. As a bit of information goes around the Cambridge ring, it is examined by all the stations connected to the ring, regardless of whether they are transmitting, receiving or idle. That examination takes time d/b per station, where b is the bandwidth and d is some small integer such as 3 or 4. These two parameters, together with the number of stations connected (including the central controller), N, the length of the wire, L, the speed of signal propagation in the wire, c, and the number of bits per slot, w, determine the physical characteristics of the network.

Give expressions for the ring delay, τ, bit capacity, B, number of slots, S and wasted bit space, $B - Sw$, in terms of the above parameters.

2. Implement the solution of equations (18), using the expressions from exercise 1, and plot the average response time, W, given by (20), against the length of the wire, L. Consider the following two cases:

  (i) The number of stations, N, remains constant.
  (ii) The number of stations increases in proportion to the length of the wire, e.g. $N = \lfloor 0.1L \rfloor$.

Keep all other parameters constant.

6.3   Token ring networks

As our last example of a centrally controlled network we shall take a ring with a rather different scheduling philosophy. In contrast to the channel sharing practised by the slotted rings, this network, known as the 'token ring', allows a station to monopolise the entire channel for the duration of a transmission. Transmissions involve whole messages, which may be of variable lengths. If two or more stations wish to transmit, they do so one after the other, without interruptions.

The availability of the channel is signalled by a token which circulates along the ring in one direction. A station which takes hold of the token may start transmitting a message. The token remains at that station until the transmission is completed, after which it is released and resumes its circulation.

The token ring architecture was adopted by the IBM company for its local area network product.

Our model has N stations connected to the ring. These are numbered $1, 2, \ldots, N$, in the direction of the movement of the token, and need not be identical. Station i is characterised by its think times, which are distributed exponentially with mean $1/\gamma_i$, and message lengths, which may have a general distribution with mean $1/\mu_i$ and second moment M_{2i} $(i = 1, 2, \ldots, N)$. At any moment in time, a given station is either in think state, or has a message awaiting transmission, or (if it is the lucky one that holds the token) is transmitting its message.

The channel is modelled by a single circulating server. That server represented by the token, moves instantaneously from station i to station $(i \bmod N) + 1$ and, whenever it finds a waiting message, stops to service it, (see figure 6.9).

Figure 6.9

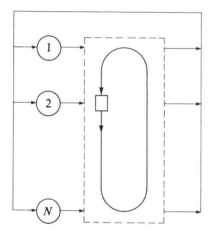

The performance measure of interest is the steady-state average response time for station i, W_i $(i = 1, 2, \ldots, N)$, defined as the average interval between the submission of a message by that station, and the completion of its transmission. Suppose that, when station i submits a message, the server is busy at station j. Then the response time of the new message consists of (a) the remaining service time at station j, (b) the time it takes the server to get from station j to station i and (c) the service time at station i. Denote the averages of components (a) and (b) by R_j and D_{ji}, respectively. The average service time at station i is of course $1/\mu_i$, and that is also the response time if, at the moment of arrival, the server is idle.

Thus, if p_{ij} is the probability that when an arrival occurs at station i the server is busy at station j $(j \neq i)$, we can write

$$W_i = \sum_{\substack{j=1 \\ j \neq i}}^{N} p_{ij}(R_j + D_{ji}) + \frac{1}{\mu_i}; \quad i = 1, 2, \ldots, N. \tag{25}$$

Next, we seek to estimate the quantities of p_{ij}, R_j and D_{ji}. The first and most obvious step is to use as R_j the residual life (see section 2.1) of the station j service time. Expression 2.(11) gives

$$R_j = \frac{\mu_j M_{2j}}{2}; \quad j = 1, 2, \ldots, N. \tag{26}$$

The approximation involved in (26) is to treat the station i arrivals as random observers, whereas in fact they may be influenced by the system state.

In order to estimate the probability p_{ij}, we shall need the fraction of time that the server spends servicing a message at station j. Note that station j goes through repetitive cycles consisting of a think period followed by a message response time. The average duration of such a cycle is $(1/\gamma_j) + W_j$. During each cycle, the station spends an average of $1/\mu_j$ actually transmitting. Therefore, the fraction of time that station j spends transmitting, i.e. the probability that the server is busy at station j, is equal to $(1/\mu_j)/[(1/\gamma_j) + W_j]$ $(j = 1, 2, \ldots, N)$. Similarly, the probability that the server is *not* busy at station i is equal to $1 - (1/\mu_i)/[(1/\gamma_i) + W_i]$. Since a message submitted by station i cannot find the server at that station, the appropriate estimate for p_{ij} is the conditional probability that the server is at station j, given that it is not at station i:

$$p_{ij} = \frac{\dfrac{\gamma_j}{\mu_j(1 + \gamma_j W_j)}}{1 - \dfrac{\gamma_i}{\mu_i(1 + \gamma_i W_i)}}; \quad i \neq j = 1, 2, \ldots, N. \tag{27}$$

Once again, this is an approximation based on treating the arrival of a message at station i as a random observation point.

The remaining unknown quantity appearing in (25) is the average time it takes the server to get from station j to station i. Denote the set of stations through which the server passes on the way from j to i by $[j, i]$. That set consists of

$$[j, i] = \begin{cases} \{j + 1, j + 2, \ldots, i - 1\} & \text{if } j < i \\ \{j + 1, \ldots, N, 1, \ldots, i - 1\} & \text{if } j > i \end{cases}. \tag{28}$$

Now, during its think-transmit cycle, station k spends an average of $W_k - (1/\mu_k)$ waiting. Hence, the fraction of time that there is a waiting message at station k is $[W_k - (1/\mu_k)]/[(1/\gamma_k) + W_k]$. Assume that this is also the probability that, when the server gets to station k, it finds a waiting message there. In other words, regard the circulating server as a random observer of the stations that it visits. With this approximation, D_{ji} can be expressed as

$$D_{ji} = \sum_{k \in [j,i]} \frac{W_k - \dfrac{1}{\mu_k}}{\dfrac{1}{\gamma_k} + W_k} \frac{1}{\mu_k}; \quad i \neq j = 1, 2, \ldots, N. \tag{29}$$

Substituting (26), (27) and (29) into (25), we get a set of equations giving the response times $W_1, W_2, \ldots, W_N$ in terms of themselves (and the known parameters). That is, if we introduce the vector $W = (W_1, W_2, \ldots, W_N)$, we have a fixed-point equation of the form

$$W = \phi(W), \tag{30}$$

where ϕ is the vector-valued function whose ith element is the right-hand side of the ith equation in (25). This equation can be solved by iteration, starting with some initial values of the response times, such as $W_i = 1/\mu_i$ $(i = 1, 2, \ldots, N)$.

Let us recapitulate the essential features of slotted and token rings. At the hardware level, a slotted ring behaves like a state-dependent processor-shared server. As soon as a message arrives, its packets start being transmitted, but at a rate which depends on the number of other messages sharing the ring. That rate depends also on the slot release policy: release at the destination station, if acceptable, is more efficient by about a factor of 2. On the other hand, the average response time of a message, both unconditional and conditioned upon its length, does not depend on the shape of the message length distribution. This is a consequence of the properties of processor-sharing (see section 4.4).

In a token ring, the ability to transmit several messages in parallel is abandoned in return for a more efficient use of the available bandwidth. Addressing information is included only once per message, instead of in every packet. Also, once started, a transmission utilises the full bit carrying capacity of the ring, rather than wasting some of it in slots which travel empty for parts of their journeys. However, transmissions are now carried out in sequence, and so there is waiting. Moreover, the shapes of the message length distributions have an influence on performance. The expression (26) for the average residual service time of the service in progress includes the second moment of the corresponding distribution. When there is much variability in message lengths, the average residual service time may be larger than the average service time (see section 2.1). In that respect, the token ring is similar to the $M/G/1$ queue (section 3.2). Performance may be bad not because the load is high, but because the variability of the demand is high.

Exercises

1. If the N stations are statistically identical, then the token ring model can be analysed exactly. Since the service time distributions are all the same, the circulating server can be replaced by a stationary one, serving messages which join a queue in FIFO order, without affecting the average response time W. One can ignore the identity of the messages submitted for transmission and concentrate only on their number.

Consider the system at consecutive service completion instants and denote by π_i the steady-state probability that, just after such an instant, there are i messages at the server ($i = 0, 1, \ldots, N - 1$). Let also ξ_{ij} be the probability that j messages arrive during a service time, given that at the start of the service there were i messages present. Show that the following balance equations are satisfied:

$$\pi_i = (\pi_0 + \pi_1)\xi_{1i} + \sum_{j=2}^{i+1} \pi_j \xi_{j,i+1-j}; \quad i = 0, 1, \ldots, N - 1, \tag{31}$$

where the second term in the right-hand side is 0 by definition of $i = 0$. Also, show that the probabilities ξ_{ij} are given by

$$\xi_{ij} = \int_0^\infty \binom{N-i}{j}(1 - e^{-\gamma x})^j (e^{-\gamma x})^{N-i-j} f(x)\,dx; \quad j \leqslant N - i, \tag{32}$$

where $1/\gamma$ is the average think time and $f(x)$ is the density function of the service time (the same for all stations). Finally, find the message throughput by arguing that the average interval between two consecutive departures is equal to the average service time with probability $1 - \pi_0$ (i.e. if the server was not idle after the last departure) and to the sum of an average service time plus an average idle period with probability π_0. The average response time is then obtained from the familiar formula (20).

2. In relations (27) and (29), the quantities p_{ij} and D_{ij} were estimated in terms of the average response times, W_j, for $j \neq i$. An alternative approach would be to relate those quantities to the response times in a system with $N - 1$ stations (namely a system where station i does not exist). This can be justified by saying that station i submits a message only after it has been out of action for a period of time.

Follow the alternative approach and obtain a set of recurrence relations for W_i, with respect to the number of stations.

Literature

This chapter is based mainly on journal articles and technical reports, plus some original material. The discussion of the achievable throughput in broadcast networks follows, at least in spirit, that in Kleinrock and Tobagi [21]. The results on the time to instability of the slotted ALOHA channel came from Greenberg and Weiss [13]. A more detailed treatment of slotted and token rings can be found in King and Mitrani [17], Mitrani, Falconer and Adams [27] and King, Mitrani and Plateau [18].

For a general and extensive coverage of the topic of computer networks the interested reader is directed to a book such as Tanenbaum [32].

Bibliography

[1] Bruell, S. C. and Balbo, G., *Computational Algorithms for Closed Queueing Networks*, North-Holland, 1980.

[2] Cinlar, E., *Introduction to Stochastic Processes*, Prentice-Hall, 1975.

[3] Coffman, E. G., Jr and Denning, P. J., *Operating Systems Theory*, Prentice-Hall, 1973.

[4] Cohen, J. W., *The Single Server Queue*, Wiley-Interscience, 1969.

[5] Conway, R. W., Maxwell, W. L. and Miller, L. W., *Theory of Scheduling*, Addison-Wesley, 1967.

[6] Courtois, P. J., *Decomposability: Queueing and Computer System Applications*, Academic Press, 1977.

[7] Cox, D. R., *Renewal Theory*, Methuen; John Wiley, 1962.

[8] Doob, J. L., *Stochastic Processes*, John Wiley, 1953.

[9] Feller, W., *An Introduction to Probability Theory and Its Applications*, volume 1, John Wiley, 1968.

[10] Foster, F. G., *Stochastic Processes, Proceedings of the IFORS Conference*, Dublin, 1972.

[11] Gelenbe, E. and Mitrani, I., *Analysis and Synthesis of Computer Systems*, Academic Press, 1980.

[12] Gittins, J. C. and Jones, D. M., A Dynamic Allocation Index for the Sequential Design of Experiments, in *Progress in Statistics* (ed. J. Gani), North-Holland, 1974.

[13] Greenberg, A. G. and Weiss, A., An Analysis of ALOHA System Via Large Deviations, AT&T Bell Laboratories Technical Report, 1986.

[14] Jackson, J. R., Networks of Waiting Lines, *Opns Res.*, **5**, 4, 1957.

[15] Jaiswal, N. K., *Priority Queues*, Academic Press, 1968.

[16] Kelly, F. P., *Reversibility and Stochastic Networks*, John Wiley, 1979.

[17] King, P. J. B. and Mitrani, I., Modelling the Cambridge Ring, *Perf. Eval. Review*, **11**, 4, 1982.

[18] King, P. J. B., Mitrani, I. and Plateau, B., Modelisation d'Un Reseau Local Avec Jeton, Université Paris-Sud ISEM Technical Report, 1983.

[19] Kingman, J. F. C. and Taylor, S. J., *Introduction to Measure and Probability*, Cambridge University Press, 1966.

[20] Kleinrock, L., *Queueing Systems*, volumes 1 and 2, John Wiley, 1975, 1976.

[21] Kleinrock, L. and Tobagi, F. A., Packet Switching in Radio Channels: Part 1 – Carrier Sense Multiple Access Modes and Their Throughput-Delay Characteristics, *IEEE Trans. Comm.*, **23**, 12, 1975.

[22] Kobayashi, H., *Modelling and Analysis*, Addison-Wesley, 1978.

[23] Lavenberg, S. S., *Computer Performance Modelling Handbook*, Academic Press, 1982.

[24] Lazowska, E. D., Zahorjan, J., Graham, G. S. and Sevcik, K. C., *Quantitative System Performance*, Prentice-Hall, 1984.

[25] Little, J. D. C., A Proof for the Queueing Formula $L = \lambda W$, *Opns Res.*, **9**, 1961.

[26] Loeve, M., *Probability Theory*, Van Nostrand, 1955.

[27] Mitrani, I., Falconer, R. M. and Adams, J., A Model of the Orwell Ring Protocol, *Proceedings of the International Teletraffic Symposium*, Amsterdam, 1986.

[28] Parzen, E., *Stochastic Processes*, Holden-Day, 1962.

[29] Saaty, T. L., *Elements of Queueing Theory and its Applications*, McGraw-Hill, 1961.

[30] Sauer, C. H. and Chandy, K. M., *Computer Systems Performance Modelling*, Prentice-Hall, 1981.

[31] Sevcik, K. C., A Proof of the Optimality of Smallest Rank Scheduling, *Journal of the Association for Computing Machinery*, **21**, 1974.

[32] Tanenbaum, A. S., *Computer Networks*, Prentice-Hall, 1981.

[33] Whittle, P., *Probability*, Penguin, 1970.

[34] Whittle, P., *Optimization Over Time, Dynamic Programming and Stochastic Control*, John Wiley, 1982.

Index